Encyclopaedia of

Tools and Techniques in Biochemistry and Molecular Biology

Encyclopaedia of

Tools and Techniques in Biochemistry and Molecular Biology

Volume 2

Editor:
Rashmi Rana

AGROTECH PRESS

JAIPUR ▪ NEW DELHI

AGROTECH PRESS

Regd. Off. "Bony Residency", Gate No. 2
Opp. Tilak Public School, Vishweswariya Nagar,
Gopalpura Road, Jaipur - 302018, India
Ph: 0141 - 2761280, Telefax: 2761381
E-mail : agrotechpress@gmail.com

Delhi Office: 102, 1st Floor, Satyam House,
4327/3, Ansari Road, Darya Ganj,
New Delhi - 110002, India
Ph: 011-45652440
visit us: www.agrotechpress.com

Distributed By
KEYA INTERNATIONAL PRESS LLP
E-mail: keyainternationalpressllp@gmail.com

Published in India, 2025

ISBN : 9789384568245 (4 Vols. Set)

Typeset by Thomson Press

Printed and bound in India by Thomson Press Ltd.,
New Delhi

Contents

Preface

Molecular biology is the branch of biology that deals with the molecular basis of biological activity. This field overlaps with other areas of biology and chemistry, particularly genetics and biochemistry. Molecular biology chiefly concerns itself with understanding the interactions between the various systems of a cell, including the interactions between the different types of DNA, RNA and protein biosynthesis as well as learning how these interactions are regulated.

Since the late 1950s and early 1960s, molecular biologists have learned to characterize, isolate, and manipulate the molecular components of cells and organisms. These components include DNA, the repository of genetic information; RNA, a close relative of DNA whose functions range from serving as a temporary working copy of DNA to actual structural and enzymatic functions as well as a functional and structural part of the translational apparatus; and proteins, the major structural and enzymatic type of molecule in cells. One of the most basic techniques of molecular biology to study protein function is expression cloning. In this technique, DNA coding for a protein of interest is cloned into a plasmid. A vector has 3 distinctive features: an origin of replication, a multiple cloning site (MCS), and a selective marker (usually antibiotic resistance). The origin of replication will have promoter regions upstream from the replication/transcription start site. The polymerase chain reaction is an extremely versatile technique for copying DNA. In brief, PCR allows a single DNA sequence to be copied (millions of times), or altered in predetermined ways. For example, PCR can be used to introduce restriction enzyme sites, or to mutate (change) particular bases of DNA, the latter is a method referred to as "Quick change". PCR can also be used to determine whether a particular DNA fragment is found in a cDNA library. PCR has many variations, like reverse transcription PCR for amplification of RNA, and, more recently, real-time PCR which allow for quantitative measurement of DNA or RNA molecules. Gel electrophoresis is one of the principal tools of molecular biology. The basic principle is that DNA, RNA, and proteins can all be separated by means of an electric field. In agarose gel electrophoresis, DNA and RNA can be separated on the basis of size by running the DNA through an agarose gel. Proteins can be separated on the basis of size by using an SDS-PAGE gel, or on the basis of size and their electric charge by using what is known as a 2D gel electrophoresis. The terms *northern*, *western* and *eastern* blotting are derived from what initially was a molecular biology joke that played on the term *Southern blotting*, after the technique described by Edwin Southern for the hybridisation of blotted DNA. Patricia Thomas, developer of the RNA blot which then became known as the *northern blot* actually didn't use the term. Further combinations of these techniques produced such terms as *southwesterns* (protein-DNA hybridizations), *northwesterns* (to detect protein-RNA interactions) and *farwesterns* (protein-protein interactions), all of which are presently found in the literature. The northern blot is used to study the expression patterns of a specific type of RNA molecule as relative comparison among a set of different samples of RNA. It is essentially a combination of denaturing RNA gel electrophoresis, and a blot. In this process RNA is separated based on size and is then transferred to a membrane that is then probed with a labelled complement of a sequence of interest. The results may be visualized through a variety of ways depending on the label used; however, most result in the revelation of bands representing the sizes of the RNA detected in sample. The intensity of these bands is related to the amount of the target RNA in the samples analysed. The procedure is commonly used to study when and how much gene expression is occurring by measuring how much of that RNA is present in different samples. It is one of the most basic tools for determining at what time, and under what conditions, certain genes are expressed in living tissues. While molecular biology was established in the 1930s, the term was coined by Warren Weaver in 1938. Warren was the director of Natural Sciences for the Rockefeller Foundation at the time and believed that biology was about to undergo a period of significant change given recent advances in fields such as X-ray crystallography. He therefore channelled significant amounts of money into biological fields. Clinical research and medical therapies arising from molecular biology are partly covered under gene therapy. The use of molecular biology or molecular cell biology approaches in medicine is now called molecular medicine. Molecular biology also plays important role in understanding formations, actions, regulations of various parts of cells which can be used efficiently for targeting new drugs, diagnosis of disease, physiology of Cell.

The present book chapters contain first hands-on information on methods and protocols in a simplified manner which is very easy to learn and perform. Further, methods and protocols constitute a gold-standard reference for today's scientists who wish to develop and hone their molecular biology skills towards the discovery of new biological relationships. This book has been written keeping in mind the requirements of graduate/postgraduate students and research scholars.

—Editor

1

Regulation of Gene Expression

Regulation of gene expression refers to the control of the amount and timing of appearance of the functional product of a gene. Control of expression is vital to allow a cell to produce the gene products it needs when it needs them; in turn this gives cells the flexibility to adapt to a variable environment, external signals, damage to the cell, etc. Some simple examples of where gene expression is important are:

- Control of insulin expression so it gives a signal for blood glucose regulation
- X chromosome inactivation in female mammals to prevent an "overdose" of the genes it contains.
- Cyclin expression levels control progression through the eukaryotic cell cycle

More generally gene regulation gives the cell control over all structure and function, and is the basis for cellular differentiation, morphogenesis and the versatility and adaptability of any organism. Any step of gene expression may be modulated, from the DNA-RNA transcription step to post-translational modification of a protein.

The stability of the final gene product, whether it is RNA or protein, also contributes to the expression level of the gene - an unstable product results in a low expression level. In general gene expression is regulated through changes in the number and type of interactions between molecules that collectively influence transcription of DNA and translation of RNA.

Numerous terms are used to describe types of genes depending on how they are regulated, these include:

- A constitutive gene is a gene that is transcribed continually compared to a facultative gene which is only transcribed when needed.
- A housekeeping gene is typically a constitutive gene that is transcribed at a relatively constant level. The housekeeping gene's products are typically needed for maintenance of the cell. It is generally assumed that their expression is unaffected by experimental conditions. Examples include actin, GAPDH and ubiquitin.
- A facultative gene is a gene which is only transcribed when needed compared to a constitutive gene.
- An inducible gene is a gene whose expression is either responsive to environmental change or dependent on the position in the cell cycle.

Transcriptional Regulation

Regulation of transcription can be broken down into three main routes of influence; genetic (direct interaction of a control factor with the gene), modulation (interaction of a control factor with the transcription machinery) and epigenetic (non-sequence changes in DNA structure which influence transcription).

Direct interaction with DNA is the simplest and the most direct method by which a protein can change transcription levels. Genes often have several protein binding sites around the coding region with the specific function of regulating transcription. There are many classes of regulatory DNA binding sites known as enhancers, insulators, repressors and silencers. The mechanisms for regulating transcription are very varied, from blocking key binding sites on the DNA for RNA polymerase to acting as an activator and promoting transcription by assisting RNA polymerase binding.

The activity of transcription factors is further modulated by intracellular signals causing protein post-translational modification including phosphorylated, acetylated, or glycosylated. These changes influence a transcription factor's ability to bind, directly or indirectly, to promoter DNA, to recruit RNA polymerase, or to favor elongation of a newly synthetized RNA molecule.

The nuclear membrane in eukaryotes allows further regulation of transcription factors by the duration of their presence in the nucleus which is regulated by reversible changes in their structure and by binding of other proteins. Environmental stimuli or endocrine signals may cause modification of regulatory proteins eliciting cascades of intracellular signals, which result in regulation of gene expression.

More recently it has become apparent that there is a huge influence of non-DNA-sequence specific effects on translation. These effects are referred to as epigenetic and involve the higher order structure of DNA, non-sequence specific DNA binding proteins and chemical modification of DNA. In general epigenetic effects alter the accessibility of DNA to proteins and so modulate transcription.

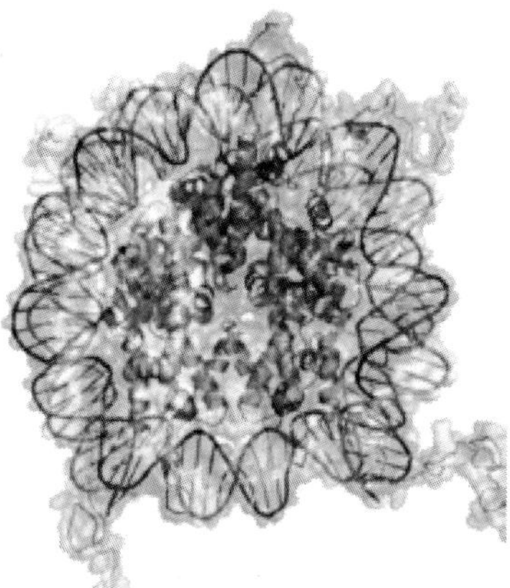

In eukaryotes, DNA is organized in form of nucleosomes. Note how the DNA (blue and green) is tightly wrapped around the protein core made of histone octamer (ribbon coils), restricting access to the DNA. From PDB 1KX5.

DNA methylation is a widespread mechanism for epigenetic influence on gene expression and is seen in bacteria and eukaryotes and has roles in heritable transcription silencing and transcription regulation. In eukaryotes the structure of chromatin, controlled by the histone code, regulates access to DNA with significant impacts on the expression of genes in euchromatin and heterochromatin areas.

Regulated Stages of Gene Expression

Any step of gene expression may be modulated, from the DNA-RNA transcription step to post-translational modification of a protein. The following is a list of stages where gene expression is regulated, the most extensively utilised point is Transcription Initiation:

- Chromatin domains
- Transcription
- Post-transcriptional modification
- RNA transport
- Translation
- mRNA degradation

Modification of DNA

In eukaryotes, the accessibility of large regions of DNA can depend on its chromatin structure, which can be altered as a result of histone modifications directed by DNA methylation, ncRNA, or DNA-binding protein. Hence these modifications may up or down regulate the expression of gene. Certain of these modifications that regulate gene expression are inheritable and are referred to as epigenetic regulation.

Chemical

Methylation of DNA is a common method of gene silencing. DNA is typically methylated by methyltransferase enzymes on cytosine nucleotides in a CpG dinucleotide sequence (also called "CpG islands" when densely clustered). Analysis of the pattern of methylation in a given region of DNA (which can be a promoter) can be achieved through a method called bisulfite mapping. Methylated cytosine residues are unchanged by the treatment, whereas unmethylated ones are changed to uracil. The differences are analyzed by DNA sequencing or by methods developed to quantify SNPs, such as Pyrosequencing (Biotage) or MassArray (Sequenom), measuring the relative amounts of C/T at the CG dinucleotide. Abnormal methylation patterns are thought to be involved in oncogenesis.

Structural

Transcription of DNA is dictated by its structure. In general, the density of its packing is indicative of the frequency of transcription. Octameric protein complexes called nucleosomes are responsible for the amount of supercoiling of DNA, and these complexes can be temporarily modified by processes such as phosphorylation or more permanently modified by processes such as methylation. Such modifications are considered to be responsible for more or less permanent changes in gene expression levels. Histone acetylation is also an important process in transcription. Histone acetyltransferase enzymes (HATs) such as CREB-binding protein also dissociate the DNA from the histone complex, allowing transcription to proceed. Often, DNA methylation and histone deacetylation work together in gene silencing. The combination of the two seems to be a signal for DNA to be packed more densely, lowering gene expression.

Regulation of Transcription

Regulation of transcription controls when transcription occurs and how much RNA is created. Transcription of a gene by RNA polymerase can be regulated by at least five mechanisms:

- Specificity factors alter the specificity of RNA polymerase for a given promoter or set of promoters, making it more or less likely to bind to them (i.e., sigma factors used in prokaryotic transcription).

- Repressors bind to non-coding sequences on the DNA strand that are close to or overlapping the promoter region, impeding RNA polymerase's progress along the strand, thus impeding the expression of the gene.
- General transcription factors position RNA polymerase at the start of a protein-coding sequence and then release the polymerase to transcribe the mRNA.
- Activators enhance the interaction between RNA polymerase and a particular promoter, encouraging the expression of the gene. Activators do this by increasing the attraction of RNA polymerase for the promoter, through interactions with subunits of the RNA polymerase or indirectly by changing the structure of the DNA.
- Enhancers are sites on the DNA helix that are bound to by activators in order to loop the DNA bringing a specific promoter to the initiation complex. Enhancers are much more common in eukaryote than prokaryotes, where only a few examples exist (to date).

Post-transcriptional Regulation

After the DNA is transcribed and mRNA is formed, there must be some sort of regulation on how much the mRNA is translated into proteins. Cells do this by modulating the capping, splicing, addition of a Poly(A) Tail, the sequence-specific nuclear export rates, and, in several contexts, sequestration of the RNA transcript. These processes occur in eukaryotes but not in prokaryotes. This modulation is a result of a protein or transcript that, in turn, is regulated and may have an affinity for certain sequences.

Regulation of Translation

The translation of mRNA can also be controlled by a number of mechanisms, mostly at the level of initiation. Recruitment of the small ribosomal subunit can indeed be modulated by mRNA secondary structure, antisense RNA binding, or protein binding. In both prokaryotes and eukaryotes, a large number of RNA binding proteins exist, which often are directed to their target sequence by the secondary structure of the transcript, which may change depending on certain conditions, such as temperature or presence of a ligand (aptamer). Some transcripts act as ribozymes and self-regulate their expression.

Examples of Gene Regulation

- Enzyme induction is a process in which a molecule (e.g., a drug) induces (i.e., initiates or enhances) the expression of an enzyme.
- The induction of heat shock proteins in the fruit fly Drosophila melanogaster.
- The Lac operon is an interesting example of how gene expression can be regulated.

- Viruses, despite having only a few genes, possess mechanisms to regulate their gene expression, typically into an early and late phase, using collinear systems regulated by anti-terminators (lambda phage) or splicing modulators (HIV).

Developmental Biology

A large number of studied regulatory systems come from developmental biology. Examples include:

- The colinearity of the Hox gene cluster with their nested antero-posterior patterning
- It has been speculated that pattern generation of the hand (digits - interdigits) The gradient of Sonic hedgehog (secreted inducing factor) from the zone of polarizing activity in the limb, which creates a gradient of active Gli3, which activates Gremlin, which inhibits BMPs also secreted in the limb, resulting in the formation of an alternating pattern of activity as a result of this reaction-diffusion system.
- Somitogenesis is the creation of segments (somites) from a uniform tissue (Pre-somitic Mesoderm, PSM). They are formed sequentially from anterior to posterior. This is achieved in amniotes possibly by means of two opposing gradients, Retinoic acid in the anterior (wavefront) and Wnt and Fgf in the posterior, coupled to an oscillating pattern (segmentation clock) composed of FGF + Notch and Wnt in antiphase.
- Sex determination in the soma of a Drosophila requires the sensing of the ratio of autosomal genes to sex chromosome-encoded genes, which results in the production of sexless splicing factor in females, resulting in the female isoform of doublesex.

Morphogen

A morphogen is a substance governing the pattern of tissue development, and the positions of the various specialized cell types within a tissue. It spreads from a localized source and forms a concentration gradient across a developing tissue . In developmental biology a morphogen is rigorously used to mean a signaling molecule that acts directly on cells (not through serial induction) to produce specific cellular responses dependent on morphogen concentration .

Well-known morphogens include: Decapentaplegic / Transforming growth factor beta, Hedgehog / Sonic Hedgehog, Wingless / Wnt, Epidermal growth factor, Fibroblast growth factor, and Retinoic acid.

Morphogens are defined conceptually, not chemically, so simple compounds such as retinoic acid (the active metabolite of retinol or vitamin A) may also act as morphogens

Introduction

During early development, morphogen gradients generate different cell types in distinct spatial order. The morphogen provides spatial information by forming a concentration gradient that subdivides a field of cells by inducing or maintaining the expression of different target genes at distinct concentration thresholds. Thus, cells far from the source of the morphogen will receive low levels of morphogen and express only low-threshold target genes.

In contrast, cells close to the source of morphogen will receive high levels of morphogen and will express both low- and high-threshold target genes. Distinct cell types emerge as a consequence of the different combination of target gene expression. In this way, the field of cells is subdivided into different types according to their position relative to the source of the morphogen. This is a general mechanism by which cell type diversity can be generated in animal development. Some of the earliest and best-studied morphogens are transcription factors that diffuse within early Drosophila melanogaster (fruit fly) embryos. However, most morphogens are secreted proteins that signal between cells.

Fruit Fly

Drosophila melanogaster has an unusual developmental system, in which the first thirteen cell divisions of the embryo occur within a syncytium prior to cellularization. Essentially the embryo remains a single cell with over 8000 nuclei evenly spaced near the membrane until the fourteenth cell division, when independent membranes furrow between the nuclei, separating them into independent cells.

As a result, in fly embryos transcription factors such as Bicoid or Hunchback can act as morphogens because they can freely diffuse between nuclei to produce smooth gradients of concentration without relying on specialized intercellular signalling mechanisms. Although there is some evidence that homeobox transcription factors similar to these can pass directly through cell membranes, this mechanism is not believed to contribute greatly to morphogenesis in cellularized systems.

In most developmental systems, such as human embryos or later Drosophila development, syncytia occur only rarely (such as in skeletal muscle), and morphogens are generally secreted signalling proteins. These proteins bind to the extracellular domains of transmembrane receptor proteins, which use an elaborate process of signal transduction to communicate the level of morphogen to the nucleus. The nuclear targets of signal transduction pathways are usually transcription factors, whose activity is regulated in a manner that reflects the level of morphogen received at the cell surface. Thus, secreted morphogens act to generate gradients of transcription factor activity just like those that are generated in the syncitial Drosophila embryo.

Discrete target genes respond to different thresholds of morphogen activity. The expression of target genes is controlled by segments of DNA called 'enhancers' to which transcription factors bind directly. Once bound, the transcription factor then stimulates or inhibits the transcription of the gene and thus controls the level of expression of the gene product (usually a protein). 'Low-threshold' target genes require only low levels of morphogen activity to be regulated and feature enhancers that contain many high-affinity binding sites for the transcription factor. 'High-threshold' target genes have relatively fewer binding sites or low-affinity binding sites that require much greater levels of transcription factor activity to be regulated.

Thus, the general mechanism by which morphogens subdivide tissues into patterns of distinct cell types is well understood. However, morphogens often have additional activities such as controlling the growth of the tissue or orienting the polarity of cells within it (for example, the hairs on your forearm point in one direction). These activities of morphogens are much less understood and are the subject of current research efforts in the field of developmental biology.

Discovery

The morphogen idea has a long history in developmental biology, dating back to the work of the pioneering Drosophila geneticist, Thomas Hunt Morgan, in the early 20th century. However, it was Lewis Wolpert who refined the morphogen concept in the 1960s with his famous French flag model which described how morphogen could subdivide a tissue into domains of different target gene expression (corresponding to the colours of the French flag).

This model was championed by the leading Drosophila biologist, Peter Lawrence. Christiane Nusslein-Volhard identified the first morphogen, Bicoid, one of the transcription factors present in a gradient in the Drosophila syncitial embryo. Two labs, that of Gary Struhl and that of Stephen Cohen, then demonstrated that a secreted signalling protein, Decapentaplegic (the Drosophila homologue of Transforming Growth Factor Beta), acted as a morphogen during later stages of Drosophila development.

Subsequent studies of the development of many different animals has confirmed the widespread importance of morphogens in governing animal development.

Chapter. Gene Regulatory Network

A gene regulatory network or genetic regulatory network (GRN) is a collection of DNA segments in a cell which interact with each other indirectly (through their RNA and protein expression products) and with other substances in the cell, thereby governing the rates at which genes in the network are transcribed into mRNA. In general, each mRNA molecule goes on to make a specific protein (or set of proteins). In some cases this protein will be structural, and will accumulate at the cell membrane or within the cell to give it particular structural properties.

In other cases the protein will be an enzyme, i.e., a micro-machine that catalyses a certain reaction, such as the breakdown of a food source or toxin. Some proteins though serve only to activate other genes, and these are the transcription factors that are the main players in regulatory networks or cascades. By binding to the promoter region at the start of other genes they turn them on, initiating the production of another protein, and so on. Some transcription factors are inhibitory.

In single-celled organisms, regulatory networks respond to the external environment, optimising the cell at a given time for survival in this environment. Thus a yeast cell, finding itself in a sugar solution, will turn on genes to make enzymes that process the sugar to alcohol. This process, which we associate with wine-making, is how the yeast cell makes its living, gaining energy to multiply, which under normal circumstances would enhance its survival prospects.

In multicellular animals the same principle has been put in the service of gene cascades that control body-shape. Each time a cell divides, two cells result which, although they contain the same genome in full, can differ in which genes are turned on and making proteins. Sometimes a 'self-sustaining feedback loop' ensures that a cell maintains its identity and passes it on. Less understood is the mechanism of epigenetics by which chromatin modification may provide cellular memory by blocking or allowing transcription. A major feature of multicellular animals is the use of morphogen gradients, which in effect provide a positioning system that tells a cell where in the body it is, and hence what sort of cell to become. A gene that is turned on in one cell may make a product that leaves the cell and diffuses through adjacent cells, entering them and turning on genes only when it is present above a certain threshold level.

These cells are thus induced into a new fate, and may even generate other morphogens that signal back to the original cell. Over longer distances morphogens may use the active process of signal transduction. Such signalling controls embryogenesis, the building of a body plan from scratch through a series of sequential steps. They also control maintain adult bodies through feedback processes, and the loss of such feedback because of a mutation can be responsible for the cell proliferation that is seen in cancer. In parallel with this process of building structure, the gene cascade turns on genes that make structural proteins that give each cell the physical properties it needs. It has been suggested that, because biological molecular interactions are intrinsically stochastic, gene networks are the result of cellular processes and not their cause (i.e. Cellular Darwinism). However, recent experimental evidence has favored the attractor view of cell fates.

Overview

At one level, biological cells can be thought of as "partially-mixed bags" of biological chemicals – in the discussion of gene regulatory networks, these chemicals are mostly the mRNAs and proteins that arise from gene expression. These mRNA and proteins interact with each other with various degrees of specificity. Some diffuse around the cell. Others are bound to cell membranes, interacting with molecules in the environment. Still others pass through cell membranes and mediate long range signals to other cells in a multi-cellular organism. These molecules and their interactions comprise a gene regulatory network. A typical gene regulatory network looks something like this:

The nodes of this network are proteins, their corresponding mRNAs, and protein/protein complexes. Nodes that are depicted as lying along vertical lines are associated with the cell/environment interfaces, while the others are free-floating and diffusible. Implied are genes, the DNA sequences which are transcribed into the mRNAs that translate into proteins. Edges between nodes represent individual molecular reactions, the protein/protein and protein/mRNA interactions through which the products of one gene affect those of another, though the lack of experimentally obtained information often implies that some reactions are not modeled at such a fine level of detail.

These interactions can be inductive (the arrowheads), with an increase in the concentration of one leading to an increase in the other, or inhibitory (the filled circles), with an increase in one leading to a decrease in the other. A series of edges indicates a chain of such dependences, with cycles corresponding to feedback loops. The network structure is an abstraction of the system's chemical dynamics, describing the manifold ways in which one substance affects all the others to which it is connected. In practice, such GRNs are inferred from the biological literature on a given system and represent a distillation of the collective knowledge about a set of related biochemical reactions.

Genes can be viewed as nodes in the network, with input being proteins such as transcription factors, and outputs being the level of gene expression. The node itself can also be viewed as a function which can be obtained by combining basic functions upon the inputs (in the Boolean network described below these are Boolean functions, typically AND, OR, and NOT). These functions have been interpreted as performing a kind of information processing within the cell, which determines cellular behavior. The basic drivers within cells are concentrations of some proteins, which determine both spatial (location within the cell or tissue) and temporal (cell cycle or developmental stage) coordinates of the cell, as a kind of "cellular memory". The gene networks are only beginning to be understood, and it is a next step for biology to attempt to deduce the functions for each gene "node", to help understand the behavior of the system in increasing levels of complexity, from gene to signaling pathway, cell or tissue level.

Mathematical models of GRNs have been developed to capture the behavior of the system being modeled, and in some cases generate predictions corresponding with experimental observations. In some other cases, models have proven to make accurate novel predictions, which can be tested experimentally, thus suggesting new approaches to explore in an experiment that sometimes wouldn't be considered in the design of the protocol of an experimental laboratory.

The most common modeling technique involves the use of coupled ordinary differential equations (ODEs). Several other promising modeling techniques have been used, including Boolean networks, Petri nets, Bayesian networks, graphical Gaussian models, Stochastic, and Process Calculi. Conversely, techniques have been proposed for generating models of GRNs that best explain a set of time series observations.

Modelling

Coupled ODEs

For an example of modelling of the cell cycle with ODEs, see cellular model.

It is common to model such a network with a set of coupled ordinary differential equations (ODEs) or stochastic ODEs, describing the reaction kinetics of the constituent parts. Suppose that our regulatory network has N nodes, and let $S_1(t), S_2(t), ..., S_N(t)$ represent the concentrations of the N corresponding substances at time t. Then the temporal evolution of the system can be described approximately by

$$\frac{dS_j}{dt} = f_j(S_1, S_2, ..., S_N)$$

where the functions f_j express the dependence of S_j on the concentrations of other substances present in the cell. The functions f_j are ultimately derived from basic principles of chemical kinetics or simple expressions derived from these e.g. Michaelis-Menten enzymatic kinetics. Hence, the functional forms of the f_j are usually chosen as low-order polynomials or Hill functions that serve as an ansatz for the real molecular dynamics. Such models are then studied using the mathematics of nonlinear dynamics. System-specific information, like reaction rate constants and sensitivities, are encoded as constant parameters.

By solving for the fixed point of the system:

$$\frac{dS_j}{dt} = 0$$

for all j, one obtains (possibly several) concentration profiles of proteins and mRNAs that are theoretically sustainable (though not necessarily stable). Steady states of kinetic equations thus correspond to potential cell types, and oscillatory solutions to the above equation to naturally cyclic cell types.

Mathematical stability of these attractors can usually be characterized by the sign of higher derivatives at critical points, and then correspond to biochemical stability of the concentration profile. Critical points and bifurcations in the equations correspond to critical cell states in which small state or parameter perturbations could switch the system between one of several stable differentiation fates.

Trajectories correspond to the unfolding of biological pathways and transients of the equations to short-term biological events. For a more mathematical discussion, see the articles on nonlinearity, dynamical systems, bifurcation theory, and chaos theory.

Boolean Network

The following example illustrates how a Boolean network can model a GRN together with its gene products (the outputs) and the substances from the environment that affect it (the inputs). Stuart Kauffman was amongst the first biologists to use the metaphor of Boolean networks to model genetic regulatory networks.

1. Each gene, each input, and each output is represented by a node in a directed graph in which there is an arrow from one node to another if and only if there is a causal link between the two nodes.
2. Each node in the graph can be in one of two states: on or off.
3. For a gene, "on" corresponds to the gene being expressed; for inputs and outputs, "on" corresponds to the substance being present.
4. Time is viewed as proceeding in discrete steps. At each step, the new state of a node is a Boolean function of the prior states of the nodes with arrows pointing towards it.

The validity of the model can be tested by comparing simulation results with time series observations.

Continuous Networks

Continuous network models of GRNs are an extension of the boolean networks described above. Nodes still represent genes and connections between them regulatory influences on gene expression. Genes in biological systems display a continuous range of activity levels and it has been argued that using a continuous representation captures several properties of gene regulatory networks not present in the Boolean model. Formally most of these approaches are similar to an artificial neural network, as inputs to a node are summed up and the result serves as input to a sigmoid function, e.g., but proteins do often control gene expression in a synergistic, i.e. non-linear, way. However there is now a continuous network model that allows grouping of inputs to a node thus realizing another level of regulation. This model is formally closer to a higher order recurrent neural network. The same model has also been used to mimic the evolution of cellular differentiation and even multicellular morphogenesis.

Stochastic Gene Networks

Recent experimental results have demonstrated that gene expression is a stochastic process. Thus, many authors are now using the stochastic formalism, after the work by. Works on single gene expression and small synthetic genetic networks, such as the genetic toggle switch of Tim Gardner and Jim Collins, provided additional experimental data on the phenotypic variability and the stochastic nature of gene expression. The first versions of stochastic models of gene expression involved only instantaneous reactions and were driven by the Gillespie algorithm.

Since some processes, such as gene transcription, involve many reactions and could not be correctly modeled as an instantaneous reaction in a single step, it was proposed to model these reactions as single step multiple delayed reactions in order to account for the time it takes for the entire process to be complete.

From here, a set of reactions were proposed that allow generating GRNs. These are then simulated using a modified version of the Gillespic algorithm, that can simulate multiple time delayed reactions (chemical reactions where each of the products is provided a time delay that determines when will it be released in the system as a "finished product").

For example, basic transcription of a gene can be represented by the following single-step reaction (RNAP is the RNA polymerase, RBS is the RNA ribosome binding site, and Pro_i is the promoter region of gene i):

$$RNAP + \mathrm{Pr}\,o_i \xrightarrow{k_I,bas} \mathrm{Pr}\,o_i\left(\tau_i^1\right) + RBS_i\left(\tau_i^1\right) + RNAp\left(\tau_I^2\right)$$

A recent work proposed a simulator (SGNSim, Stochastic Gene Networks Simulator), that can model GRNs where transcription and translation are modeled as multiple time delayed events and its dynamics is driven by a stochastic simulation algorithm (SSA) able to deal with multiple time delayed events. The time delays can be drawn from several distributions and the reaction rates from complex functions or from physical parameters. SGNSim can generate ensembles of GRNs within a set of user-defined parameters, such as topology. It can also be used to model specific GRNs and systems of chemical reactions. Genetic perturbations such as gene deletions, gene over-expression, insertions, frame shift mutations can also be modeled as well.

The GRN is created from a graph with the desired topology, imposing in-degree and out-degree distributions. Gene promoter activities are affected by other genes expression products that act as inputs, in the form of monomers or combined into multimers and set as direct or indirect. Next, each direct input is assigned to an operator site and different transcription factors can be allowed, or not, to compete for the same operator site, while indirect inputs are given a target. Finally, a function is assigned to each gene, defining the gene's response to a combination of transcription factors (promoter state). The transfer functions (that is, how genes respond to a combination of inputs) can be assigned to each combination of promoter states as desired. In other recent work, multiscale models of gene regulatory networks have been developed that focus on synthetic biology applications. Simulations have been used that model all biomolecular interactions in transcription, translation, regulation, and induction of gene regulatory networks, guiding the design of synthetic systems.

Prediction

Other work has focused on predicting the gene expression levels in a gene regulatory network. The approaches used to model gene regulatory networks have been constrained to be interpretable and, as a result, are generally simplified versions of the network. For example, Boolean networks have been used due to their simplicity and ability to handle noisy data but lose data information by having a binary representation of the genes. Also, artificial neural networks omit using a hidden layer so that they can be interpreted, losing the ability to model higher order correlations in the data. Using a model that is not constrained to be interpretable, a more accurate model can be produced. Being able to predict gene expressions more accurately provides a way to explore how drugs affect a system of genes as well as for finding which genes are interrelated in a process. This has been encouraged by the DREAM competition which promotes a competition for the best prediction algorithms. Some other recent work has used artificial neural networks with a hidden layer.

Network Connectivity

Empirical data indicate that biological gene networks are sparsely connected, and that the average number of upstream-regulators per gene is less than two. Theoretical results show that selection for robust gene networks will favor minimally complex, more sparsely connected, networks. These results suggest that a sparse, minimally connected, genetic architecture may be a fundamental design constraint shaping the evolution of gene network complexity.

Body plan

A body plan is the blueprint for the way the body of an organism is laid out. An organism's symmetry, its number of body segments and number of limbs are all aspects of its body plan. One of the key issues of developmental biology is the evolution of body plans as different as those of a starfish, a fern, or a mammal, from a common biological heritage, and in particular how radical changes in body plans have occurred over geological time.

The body plan is a key feature of an organism's morphology and, since the discovery of DNA, developmental biologists have been able to learn a lot about how genes control the development of structural features through a cascade of processes in which key genes produce morphogens, chemicals that diffuse through the body to produce a gradient that acts as a position indicator for cells, turning on other genes, some of which in turn produce other morphogens. A key discovery was the existence of groups of homeobox genes which are responsible for laying down the basic body plan in organisms.

The homeobox genes are remarkably conserved between species as diverse as the fruitfly and man, the basic segmented pattern of the worm or fruitfly being the origin of the segmented spine in man. The field of evolutionary developmental biology, which studies the genetics of morphology in detail is now a rapidly expanding one , with many of the developmental genetic cascades, particularly in the fruitfly drosophila, now catalogued in considerable detail.

Body plan is the basis for phylum, and there are 35 different basic animal body plans, corresponding to different phyla.

Origin

The evolution of body plans became inevitable with the emergence of differentiated multicellular life in the Ediacaran Era, over 600 million years ago. The most basic and successful structure, for free-moving organisms, is the "pipe" or alimentary canal. This is common even to organisms as diverse as humans and earthworms. It is essentially a passage having a mouth at one end, and a cloaca or anus at the other. The simple process of nutrient capture, digestion, and waste disposal is fundamental to the body plan of advanced, free-moving animals. Vertebra, limbs, even brains are supplementary to the pipe. Natural selection has spun off an enormous range of variations on this basic theme, but the pipe model itself remains. The basic symmetry and organization of this body plan apparently gave an ancient organism an enormous advantage at survival and reproduction, and it has been preserved in most animals ever since.

The Cambrian explosion refers to the massive increase in different body plans that took place around 530 million years ago. Fossils from this era show all sorts of odd shapes, many quite unlike anything found today. At that time it was possible for organisms to survive and make a living even though they were unrefined and unlikely, because predation had yet to evolve, along with arms races that would optimise and streamline them to occupy a particular ecological niche.

Bauplan

Bauplan (German for building plan, blueprint; plural: baupläne or bauplaene) is a closely related term in biology referring to the common new and original (homologous) properties of the members of a systematic group (taxon). It is not necessary that a bauplan precisely describes any one particular species of that group.

The concept of bauplan is employed in the studies of morphology, taxonomy, comparative physiology, evolutionary physiology, and, most notably, phylogenetics and evolution. Before the advent of genetic sequencing, the analysis of the bauplan of fossils was the primary method to determine hypothetical relationships and lineages of species, both living and extinct. The idea is that species that are closely related share more common properties, hence a more detailed bauplan. Small differences of bauplan are indicative of species belonging to a parent, child or sibling taxon.

Genetic Basis

Similarities and differences in adult shape and form, as well as the developmental pattern of embryos, provide the framework for modern taxonomic classification. These comparisons are the basis of phylogenetic systematics. Embryonic development is relatively consistent among animals with similar body plans, although similar larval forms may give rise to very different adults in some groups. The timing, pattern, and scale of developmental events determine the shape of an organism, and closely related groups are more likely to share structural and developmental similarities than those that are more distantly related. Homologous structures and developmental stages—those that are similar among related groups because they are inherited from a common ancestor—are the basis of modern biological classification.

Examples

Figure: *An early tetrapod: the ichthyostega*

The current range of body plans is far from exhausting the possible patterns for life: the Ediacaran biota appears to contain numerous species and taxa with body plans quite different from any found in currently living organisms.

The most commonly seen body plan amongst vertebrates is that of the tetrapod, which include all mammals, birds, amphibians and reptiles. Some animal groups, such as the cetaceans, bats and most birds have been modified (e.g. front limbs become wings or flippers) but nevertheless, they are still tetrapods.

The invertebrates employ a much more diverse array of body plans, such as seen in insects (six legs, three body parts and an exoskeleton), cephalopods (no skeleton, hydrostatically stiffened tentacles, primary propulsion by squeezing water out of a mantle cavity), echinoderms (fivefold radial symmetry, external skeleton, movement by hydrostatically operated tube feet) and various phyla of "worms" (tube-shaped, movement by expanding and contracting parts of the body).

The most varied collection of body forms known is found in the Burgess Shale, where fossils from a Cambrian sea show a tremendous variety of body forms that came to rise (only to later fall extinct) during the Cambrian explosion.

Fictional

One common theme in science fiction is the appearance of extraterrestrial beings, descriptions of which have ranged from being simple variants on human anatomy to beings with body plans wildly different from any found on Earth. The field of exobiology attempts to bring these and similar speculations into the realm of serious scientific investigation.

Cis-regulatory Module

Cis-regulatory module (CRM) is a stretch of DNA, usually 100-1000 DNA base pairs in length, where a number of transcription factors can bind and regulate expression of nearby genes. One cis-regulatory element can regulate several genes, and conversely, one gene can have several cis-regulatory modules.

Cis-regulatory modules are one of several types of functional regulatory elements. Regulatory elements are binding sites for transcription factors, which are involved in gene regulation. Cis-regulatory modules perform a large amount of developmental information processing. Cis-regulatory modules are non-random clusters at their specified target site that contain transcription factor binding sites.

They are labeled as cis because they are typically located on the same DNA as the genes they control as opposed to trans, which refers to effects on genes not located on the same strand or farther away, such as transcription factors. The original definition presented cis-regulatory modules as enhancers of cis-acting DNA, which increased the rate of transcription from a linked promoter. However, this definition has changed to define cis-regulatory modules as a DNA sequence with transcription factor binding sites which are clustered into modular structures, including -but not limited to- locus control regions, promoters, enhancers, silencers, boundary control elements and other modulators. Cis-regulatory modules can be divided into classes. Enhancers regulate gene expression positively. Insulators work indirectly by interacting with other nearby cis-regulatory modules. Silencers act by silencing genes. Cis-regulatory modules carry out their function by integrating the active transcription factors and the associated co-factors at a specific time and place in the cell, this information is read and an output is given.

Gene-regulation function of a Cis-regulatory Module

The design of cis-regulatory modules is such that transcription factors and epigenetic modifications serve as inputs, and the output of the module is the command given to the transcription machinery, which in turn determines the rate of gene transcription or whether it is turned on or off. There are two types of transcription factor inputs: those that determine when the target gene is to be expressed and those that serve as functional drivers, which come into play only during specific situations during development.

These inputs can come from different time points, can represent different signal ligands, or can come from different domains or lineages of cells. However, a lot still remains unknown. Additionally, the regulation of chromatin structure and nuclear organization also play a role in determining and controlling the function of cis-regulatory modules.

Thus gene-regulation functions (GRF) provide a unique characteristic of a cis-regulatory module (CRM), relating the concentrations of transcription factors (input) to the promoter activities (output). The challenge is to predict GRFs. This challenge still remains unsolved. In general, gene-regulation functions do not use Boolean logic, although in some cases the approximation of the Boolean logic is still very useful.

The Boolean Logic Assumption

Within the assumption of the Boolean logic, four principles guiding the operation of these cis-regulatory modules are : 1) The design of the cis-regulatory module determines the regulatory function. 2) In relation to development, these modules can generate both positive and negative outputs. 3) The output of each module is a product of the various operations performed on it. Common operations: "OR" logic gate- This design indicates that in an output will be given when either input is given . "AND" logic gate– In this design two different regulatory factors are necessary to make sure that a positive output results. "Toggle Switches"- This design occurs when the signal ligand is absent while the transcription factor is present; this transcription factor ends up acting as a dominant repressor. However, once the signal ligand is present the transcription factor's role as repressor is eliminated and transcription can occur. Other Boolean logic operations can occur as well, such as sequence specific transcriptional repressors, which when they bind to the cis-regulatory module lead to an output of zero. Additionally, besides influence from the different logic operations, the output of a "cis"-regulatory module will also be influenced by prior events. 4) Cis-regulatory modules must interact with other regulatory elements.

For the most part, even with the presence of functional overlap between cis-regulatory modules of a gene, the modules' inputs and outputs tend to not be the same.

Identification and Computational Prediction of CRMs

Besides experimentally determining CRMs, there are various bioinformatics algorithms for predicting them. Most algorithms try to search for significant combinations of transcription factor binding sites (DNA binding sites) in promoter sequences of co-expressed genes. More advanced methods combine the search for significant motifs with correlation in gene expression datasets between transcription factors and target genes. Both methods have been implemented, for example, in the ModuleMaster. Other programs created for the identification and prediction of "cis-regulatory modules include:

Stubb uses hidden Markov models to identify statistically significant clusters of transcription factor combinations. It also uses a second related genome to improve the prediction accuracy of the model.

Bayesian Networks use an algorithm that combines site predictions and tissue-specific expression data for transcription factors and target genes of interest. This model also uses regression trees to depict the relationship between the identified cis-regulatory module and the possible binding set of transcription factors.

CRÈME examine clusters of target sites for transcription factors of interest. This program uses a database of confirmed transcription factor binding sites that were annotated across the human genome. A search algorithm is applied to the data set to identify possible combinations of transcription factors, which have binding sites that are close to the promoter of the gene set of interest. The possible cis-regulatory modules are then statistically analyzed and the significant combinations are graphically represented

Active cis-regulatory modules in a genomic sequence have been difficult to identify. Problems in identification arise because often scientists find themselves with a small set of known transcription factors, so it makes it harder to identify statistically significant clusters of transcription factor binding sites. Additionally, high costs limit the use of large whole genome tiling arrays.

Classification of Cis-regulatory Modules

Cis-regulatory modules can be characterized by the information processing that they encode and the organization of their transcription factor binding sites. Additionally, cis-regulatory modules are also characterized by the way they affect the probability, proportion, and rate of transcription. Highly cooperative and coordinated cis-regulatory modules are classified as enhanceosomes.

The architecture and the arrangement of the transcription factor binding sites are critical because disruption of the arrangement could cancel out the function. Functional flexible cis-regulatory modules are called billboards. Their transcriptional output is the summation effect of the bound transcription factors. Enhancers affect the probability of a gene being activated, but have little or no effect on rate. Binary response model acts like an on/off switch for transcription. This model will increase or decrease the amount of cells that transcribe a gene, however it does not affect the rate of transcription. Rheostatic response model describes cis-regulatory modules as regulators of the initiation rate of transcription of its associated gene.

Mode of Action

Cis-regulatory modules can regulate their target genes over large distances, several models have been proposed to describe the way that these modules may communicate with their target gene promoter. DNA Scanning Model: In this model the transcription factor and cofactor complex form at the cis-regulatory module and then continues to move along the DNA sequence until it finds the target gene promoter. Looping Model: In this model the transcription factor binds to the cis-regulatory module, which then causes the looping of the DNA sequence and allows for the interaction with the target gene promoter. Facilitated Tracking Model: This model combines parts of the two previous models. The transcription factor-cis-regulatory module complex causes the looping of the DNA sequence slowly towards the target promoter and forms a stable looped configuration.

Cis-regulatory Module in Gene Regulatory Network

The function of a gene regulatory network depends on the architecture of the nodes , whose function is dependent on the multiple cis-regulatory modules. The layout of cis-regulatory modules can provide enough information to generate spatial and temporal patterns of gene expression. During development each domain, where each domain represents a different spatial regions of the embryo, of gene expression will be under the control of different cis-regulatory module(s). The design of regulatory modules help in producing feedback, feed forward, and cross-regulatory loops.

Genenetwork (Database)

GeneNetwork is a database and open source bioinformatics software resource for systems genetics. This resource is used to study gene regulatory networks that link DNA sequence variants to corresponding differences in gene and protein expression and to differences in traits such as health and disease risk. Data sets in GeneNetwork are typically made up of large collections of genotypes (e.g., SNPs) and phenotypes that are obtained from groups of related individuals, including human families, experimental crosses of strains of mice and rats, and organisms as diverse as Drosophila melanogaster, Arabidopsis thaliana, and barley.

The inclusion of genotypes for all individuals makes it practical to carry out web-based gene mapping to discover those regions of the genome that contribute to differences in gene expression, cell function, anatomy, physiology, and behavior among individuals.

History

GeneNetwork was created at the University of Tennessee Health Science Center, Memphis USA in 2000-2001. It was initially developed as a web-adapted version of Kenneth F.

Manly's Map Manager QT and QTX programs and was called WebQTL. Gene mapping data were incorporated for several mouse recombinant inbred strains. By early 2003, the first large Affymetrix gene expression data sets (whole mouse brain mRNA and hematopoietic stem cells) were incorporated and the system was renamed. GeneNetwork is now developed by an international group of developers and has mirror and development sites in Europe, Asia, and Australia. The production service is hosted on the Amazon Elastic Compute Cloud.

Organization and Use

GeneNetwork consists of two major components:

- Massive collections of genetic, genomic, and phenotype data for large families
- Sophisticated statistical analysis and gene mapping software that enable analysis of regulatory networks and genotype-to-phenotype relations

Four levels of data are usually obtained for each family or population:

1. DNA sequences and genotypes
2. Gene expression values using microarray, RNA-seq, or proteomic methods (molecular phenotypes)
3. Standard phenotypes of the type that are part of a typical medical record (e.g., blood chemistry, body weight)
4. Annotation files and metadata

The combined data types are housed together in a single relational database, but are conceptually organized and divided by species and family. The system is implemented as a LAMP (software bundle) stack. Code and a simplified version of the MySQL database are available at Sourceforge.net/projects/genenetwork/.

GeneNetwork is primarily used by researchers but has also been adopted successfully for undergraduate courses in genetics (see YouTube example), bioinformatics, physiology, and psychology. Researchers and students typically retrieve sets of genotypes and phenotypes from one or more families and use built-in statistical and mapping functions to explore relations among variables and to assemble networks of associations. Key steps include the analysis of these factors:

1. The range of variation of traits
2. Covariation among traits (scatterplots and correlations)
3. Architecture of larger networks of traits
4. Quantitative trait locus mapping and causal models of the linkage between sequence differences and phenotype differences

Data Sources

Massive expression data sets are submitted by researchers directly or are extracted from repositories such as National Center for Biotechnology Information Gene Expression Omnibus. A wide variety of cells and tissues are included—from single cell populations of the immune system, specific tissues (retina, prefrontal cortex), to entire systems (whole brain, lung, muscle, heart, fat, kidney, flower, even whole plant embryos). A typical data set is often based on hundreds of fully genotyped individuals and may also include biological replicates. Genotypes and phenotypes are taken from peer-reviewed papers. GeneNetwork includes annotation files for several RNA profiling platforms (Affymetrix, Illumina, and Agilent). RNA-seq data are also available for BXD recombinant inbred mice. Content and nomenclature are reviewed and edited by curators. Updates on coverage of species, families, tissues and measurement types are available at this site: .

Topics of annotation include the following:

- DNA sequence (SNPs, CNVs, indels)
- transcriptomes (arrays, RNA-seq)
- gene regulatory networks
- phenome

Tools and Features

There are tools on the site for a wide range of functions that range from simple graphical displays of variation in gene expression or other phenotypes, scatter plots of pairs of traits (Pearson or rank order), construction of both simple and complex network graphs, analysis of principal components and synthetic traits, QTL mapping using marker regression, interval mapping, and pair scans for epistatic interactions. Most functions work with up to 100 traits and several functions work with an entire transcriptome.

The database can be browsed and searched at the main search page. An on-line tutorial is available. Users can also download the primary data sets as text files, Excel, or in the case of network graphs, as SBML.

Code

GeneNetwork is an open source project released under the Affero General Public License (AGPLv3). The majority of code is written in Python, but includes modules and other code written in C and JavaScript. GeneNetwork also calls statistical procedures written in the R programming language. The source code and a compact database are available on GeneNetwork sites and at SourceForge.

Morphogen

A morphogen is a substance governing the pattern of tissue development, and the positions of the various specialized cell types within a tissue. It spreads from a localized source and forms a concentration gradient across a developing tissue . In developmental biology a morphogen is rigorously used to mean a signaling molecule that acts directly on cells (not through serial induction) to produce specific cellular responses dependent on morphogen concentration .

Well-known morphogens include: Decapentaplegic / Transforming growth factor beta, Hedgehog / Sonic Hedgehog, Wingless / Wnt, Epidermal growth factor, Fibroblast growth factor, and Retinoic acid. Morphogens are defined conceptually, not chemically, so simple compounds such as retinoic acid (the active metabolite of retinol or vitamin A) may also act as morphogens

Introduction

During early development, morphogen gradients generate different cell types in distinct spatial order. The morphogen provides spatial information by forming a concentration gradient that subdivides a field of cells by inducing or maintaining the expression of different target genes at distinct concentration thresholds. Thus, cells far from the source of the morphogen will receive low levels of morphogen and express only low-threshold target genes.

In contrast, cells close to the source of morphogen will receive high levels of morphogen and will express both low- and high-threshold target genes. Distinct cell types emerge as a consequence of the different combination of target gene expression. In this way, the field of cells is subdivided into different types according to their position relative to the source of the morphogen.

This is a general mechanism by which cell type diversity can be generated in animal development.

Some of the earliest and best-studied morphogens are transcription factors that diffuse within early Drosophila melanogaster (fruit fly) embryos. However, most morphogens are secreted proteins that signal between cells.

Fruit Fly

Drosophila melanogaster has an unusual developmental system, in which the first thirteen cell divisions of the embryo occur within a syncytium prior to cellularization. Essentially the embryo remains a single cell with over 8000 nuclei evenly spaced near the membrane until the fourteenth cell division, when independent membranes furrow between the nuclei, separating them into independent cells. As a result, in fly embryos transcription factors such as Bicoid or Hunchback can act as morphogens because they can freely diffuse between nuclei to produce smooth gradients of concentration without relying on specialized intercellular signalling mechanisms. Although there is some evidence that homeobox transcription factors similar to these can pass directly through cell membranes, this mechanism is not believed to contribute greatly to morphogenesis in cellularized systems.

In most developmental systems, such as human embryos or later Drosophila development, syncytia occur only rarely (such as in skeletal muscle), and morphogens are generally secreted signalling proteins. These proteins bind to the extracellular domains of transmembrane receptor proteins, which use an elaborate process of signal transduction to communicate the level of morphogen to the nucleus. The nuclear targets of signal transduction pathways are usually transcription factors, whose activity is regulated in a manner that reflects the level of morphogen received at the cell surface. Thus, secreted morphogens act to generate gradients of transcription factor activity just like those that are generated in the syncitial Drosophila embryo.

Discrete target genes respond to different thresholds of morphogen activity. The expression of target genes is controlled by segments of DNA called 'enhancers' to which transcription factors bind directly. Once bound, the transcription factor then stimulates or inhibits the transcription of the gene and thus controls the level of expression of the gene product (usually a protein). 'Low-threshold' target genes require only low levels of morphogen activity to be regulated and feature enhancers that contain many high-affinity binding sites for the transcription factor. 'High-threshold' target genes have relatively fewer binding sites or low-affinity binding sites that require much greater levels of transcription factor activity to be regulated.

Thus, the general mechanism by which morphogens subdivide tissues into patterns of distinct cell types is well understood. However, morphogens often have additional activities such as controlling the growth of the tissue or orienting the polarity of cells within it (for example, the hairs on your forearm point in one direction). These activities of morphogens are much less understood and are the subject of current research efforts in the field of developmental biology.

Discovery

The morphogen idea has a long history in developmental biology, dating back to the work of the pioneering Drosophila geneticist, Thomas Hunt Morgan, in the early 20th century. However, it was Lewis Wolpert who refined the morphogen concept in the 1960s with his famous French flag model which described how morphogen could subdivide a tissue into domains of different target gene expression (corresponding to the colours of the French flag).

This model was championed by the leading Drosophila biologist, Peter Lawrence. Christiane Nusslein-Volhard identified the first morphogen, Bicoid, one of the transcription factors present in a gradient in the Drosophila syncitial embryo. Two labs, that of Gary Struhl and that of Stephen Cohen, then demonstrated that a secreted signalling protein, Decapentaplegic (the Drosophila homologue of Transforming Growth Factor Beta), acted as a morphogen during later stages of Drosophila development.

Subsequent studies of the development of many different animals has confirmed the widespread importance of morphogens in governing animal development.

Operon

In genetics, an operon is a functioning unit of genomic DNA containing a cluster of genes under the control of a single regulatory signal or promoter. The genes are transcribed together into an mRNA strand and either translated together in the cytoplasm, or undergo trans-splicing to create monocistronic mRNAs that are translated separately, i.e. several strands of mRNA that each encode a single gene product.

The result of this is that the genes contained in the operon are either expressed together or not at all. Several genes must be both co-transcribed and co-regulated to define an operon.

Originally, operons were thought to exist solely in prokaryotes, but since the discovery of the first operons in eukaryotes in the early 1990s, more evidence has arisen to suggest they are more common than previously assumed.

History

The term "operon" was first proposed in a short paper in the Proceedings of the French Academy of Science in 1960 . From this paper, the so-called general theory of the operon was developed. This theory suggested that all genes are controlled by means of operons through a single feedback regulatory mechanism– repression. Later, it was discovered that the regulation of genes is a much more complicated process.

Indeed, it is not possible to talk of a general regulatory mechanism, because different operons have different mechanisms. Despite modifications, the development of the concept is considered a landmark event in the history of molecular biology. The first operon to be described was the lac operon in E. coli. The 1965 Nobel Prize in Physiology and Medicine was awarded to François Jacob, André Michel Lwoff and Jacques Monod for their discoveries concerning the operon and virus synthesis.

Overview

Operons occur primarily in prokaryotes but also in some eukaryotes, including nematodes such as C. elegans. , and Drosophila melanogaster flies. rRNA genes often exist in operons that have been found in a range of eukaryotes including chordates. An operon is made up of several structural genes arranged under a common promoter and regulated by a common operator. It is defined as a set of adjacent structural genes, plus the adjacent regulatory signals that affect transcription of the structural genes. The regulators of a given operon, including repressors, corepressors, and activators, are not necessarily coded for by that operon. The location and condition of the regulators, promoter, operator and structural DNA sequences can determine the effects of common mutations.

Operons are related to regulons, stimulons and modulons; whereas operons contain a set of genes regulated by the same operator, regulons contain a set of genes under regulation by a single regulatory protein, and stimulons contain a set of genes under regulation by a single cell stimulus. According to its authors, the term "operon" means "to operate".

As a Unit of Transcription

An operon contains one or more structural genes which are generally transcribed into one polycistronic mRNA (a single mRNA molecule that codes for more than one protein). However, the definition of an operon does not require the mRNA to be polycistronic, though in practice, it usually is. Upstream of the structural genes lies a promoter sequence which provides a site for RNA polymerase to bind and initiate transcription. Close to the promoter lies a section of DNA called an operator.

The operon may also contain regulatory genes such as a repressor gene, which codes for a regulatory protein that binds to the operator and inhibits transcription. Regulatory genes need not be part of the operon itself, but may be located elsewhere in the genome. The repressor molecule will reach the operator to block the transcription of the structural genes.

Structure

This is the general structure of an operon:

- Promoter – a nucleotide sequence that enables a gene to be transcribed. The promoter is recognized by RNA polymerase, which then initiates transcription. In RNA synthesis, promoters indicate which genes should be used for messenger RNA creation – and, by extension, control which proteins the cell manufactures.
- Operator – a segment of DNA that a regulator binds to. It is classically defined in the lac operon as a segment between the promoter and the genes of the operon. In the case of a repressor, the repressor protein physically obstructs the RNA polymerase from transcribing the genes.
- Structural genes – the genes that are co-regulated by the operon.

Regulation

Control of an operon is a type of gene regulation that enables organisms to regulate the expression of various genes depending on environmental conditions. Operon regulation can be either negative or positive by induction or repression. Negative control involves the binding of a repressor to the operator to prevent transcription.

- In negative inducible operons, a regulatory repressor protein is normally bound to the operator, which prevents the transcription of the genes on the operon. If an inducer molecule is present, it binds to the repressor and changes its conformation so that it is unable to bind to the operator. This allows for expression of the operon.

- In negative repressible operons, transcription of the operon normally takes place. Repressor proteins are produced by a regulator gene, but they are unable to bind to the operator in their normal conformation. However, certain molecules called corepressors are bound by the repressor protein, causing a conformational change to the active state. The activated repressor protein binds to the operator and prevents transcription.

Operons can also be positively controlled. With positive control, an activator protein stimulates transcription by binding to DNA (usually at a site other than the operator).

- In positive inducible operons, activator proteins are normally unable to bind to the pertinent DNA. When an inducer is bound by the activator protein, it undergoes a change in conformation so that it can bind to the DNA and activate transcription.
- In positive repressible operons, the activator proteins are normally bound to the pertinent DNA segment. However, when a corepressor is bound by the activator, it is prevented from binding the DNA. This stops activation and transcription of the system.

The Lac Operon

The lac operon of the model bacterium Escherichia coli was the first operon to be discovered and provides a typical example of operon function. It consists of three adjacent structural genes, a promoter, a terminator, and an operator. The lac operon is regulated by several factors including the availability of glucose and lactose. This is an example of the derepressible (from above: negative inducible) model.

The Trp Operon

Discovered in 1953 by Jacques Monod and colleagues, the trp operon in E. coli was the first repressible operon to be discovered. While the lac operon can be activated by a chemical (allolactose), the tryptophan (Trp) operon is inhibited by a chemical (tryptophan). This operon contains five structural genes: trp E, trp D, trp C, trp B, and trp A, which encodes tryptophan synthetase. It also contains a promoter which binds to RNA polymerase and an operator which blocks transcription when bound to the protein synthesized by the repressor gene (trp R) that binds to the operator. In the lac operon, lactose binds to the repressor protein and prevents it from repressing gene transcription, while in the trp operon, tryptophan binds to the repressor protein and enables it to repress gene transcription. Also unlike the lac operon, the trp operon contains a leader peptide and an attenuator sequence which allows for graded regulation. This is an example of the corepressible model.

Predicting the Number and Organization of Operons

The number and organization of operons has been studied most critically in E. coli. As a result, predictions can be made based on an organism's genomic sequence.

One prediction method uses the intergenic distance between reading frames as a primary predictor of the number of operons in the genome. The separation merely changes the frame and guarantees that the read through is efficient. Longer stretches exist where operons start and stop, often up to 40–50 bases.

An alternative method to predict operons is based on finding gene clusters where gene order and orientation is conserved in two or more genomes.

Operon prediction is even more accurate if the functional class of the molecules is considered. Bacteria have clustered their reading frames into units, sequestered by co-involvement in protein complexes, common pathways, or shared substrates and transporters. Thus, accurate prediction would involve all of these data, a difficult task indeed.

Pascale Cossart published in 2009 the first full map of an operon, identifying the genetic switches that operate in Listeria under different conditions.

Synexpression

Synexpression is a type of non-random eukaryotic gene organization. Genes in a synexpression group may not be physically linked, but they are involved in the same process and they are coordinately expressed. It is expected that genes that function in the same process be regulated coordinately. Synexpression groups in particular represent genes that are simultaneously up- or down-regulated, often because their gene products are required in stoichiometric amounts or are protein-complex subunits. It is likely that these gene groups share common cis- and trans-acting control elements to achieve coordinate expression.

Synexpression groups are determined mainly by analysis of expression profiles compiled by the use of DNA microarrays. The use of this technology helps researchers monitor changes in expression patterns for large numbers of genes in a given experiment. Analysis of DNA microarray expression profiles has led to the discovery of a number of genes that are tightly co-regulated. One simplified example of a synexpression group is the genes cdc6, cdc3, cdc46, and swi4 in yeast, which are all co-expressed early in the G-1 stage of the cell cycle., These genes share one common cis-regulatory element, called ECB, which serves as a binding site for the MCM1 trans-acting protein. Although these genes are not spatially clustered, co-regulation seems to be achieved via this common cis and trans control mechanism. Most synexpression groups are more complicated than the ECB group in yeast, involving a myriad of cis and trans control elements.,

The identification of synexpression groups has had an impact on the way some scientists view evolutionary change in higher eukaryotes. Since groups of genes involved in the same biological process often share one or more common control elements, it has been suggested that the differential expression of these synexpression groups in different tissues of organisms can contribute to co-evolution tissues, organs, and appendages.

Today it is commonly believed that it is not primarily the gene products themselves that evolve, but that it is the control networks for groups of genes that contribute most to the evolution of higher eukaryotes. Developmental processes provide an example of how changes in synexpression control networks could have a significant impact on an organism's capacity to evolve and adapt effectively. In animals, it is often beneficial for appendages to co-evolve, and it has been observed that fore-and hind-limbs share expression of Hox genes early in metazoan development. Thus, changes in the regulatory patterns of these genes would effect the development of both the fore- and hind-limbs, facilitating co-evolution.

Systems Biology

Systems biology is a term used to describe a number of trends in bioscience research, and a movement which draws on those trends. Proponents describe systems biology as a biology-based inter-disciplinary study field that focuses on complex interactions in biological systems, claiming that it uses a new perspective (holism instead of reduction).

Particularly from year 2000 onwards, the term is used widely in the biosciences, and in a variety of contexts. An often stated ambition of systems biology is the modeling and discovery of emergent properties, properties of a system whose theoretical description is only possible using techniques which fall under the remit of systems biology. These typically involve cell signaling networks, via long-range allostery.

Overview

Systems biology can be considered from a number of different aspects:

- As a field of study, particularly, the study of the interactions between the components of biological systems, and how these interactions give rise to the function and behavior of that system (for example, the enzymes and metabolites in a metabolic pathway).

- As a paradigm, usually defined in antithesis to the so-called reductionist paradigm (biological organisation), although fully consistent with the scientific method. The distinction between the two paradigms is referred to in these quotations:

"The reductionist approach has successfully identified most of the components and many of the interactions but, unfortunately, offers no convincing concepts or methods to understand how system properties emerge...the pluralism of causes and effects in biological networks is better addressed by observing, through quantitative measures, multiple components simultaneously and by rigorous data integration with mathematical models" Sauer et al

"Systems biology...is about putting together rather than taking apart, integration rather than reduction. It requires that we develop ways of thinking about integration that are as rigorous as our reductionist programmes, but different....It means changing our philosophy, in the full sense of the term" Denis Noble

- As a series of operational protocols used for performing research, namely a cycle composed of theory, analytic or computational modelling to propose specific testable hypotheses about a biological system, experimental validation, and then using the newly acquired quantitative description of cells or cell processes to refine the computational model or theory. Since the objective is a model of the interactions in a system, the experimental techniques that most suit systems biology are those that are system-wide and attempt to be as complete as possible. Therefore, transcriptomics, metabolomics, proteomics and high-throughput techniques are used to collect quantitative data for the construction and validation of models.
- As the application of dynamical systems theory to molecular biology.
- As a socioscientific phenomenon defined by the strategy of pursuing integration of complex data about the interactions in biological systems from diverse experimental sources using interdisciplinary tools and personnel.

This variety of viewpoints is illustrative of the fact that systems biology refers to a cluster of peripherally overlapping concepts rather than a single well-delineated field. However the term has widespread currency and popularity as of 2007, with chairs and institutes of systems biology proliferating worldwide.

History

Systems biology finds its roots in:

- the quantitative modeling of enzyme kinetics, a discipline that flourished between 1900 and 1970,

- the mathematical modeling of population growth,
- the simulations developed to study neurophysiology, and
- control theory and cybernetics.

One of the theorists who can be seen as one of the precursors of systems biology is Ludwig von Bertalanffy with his general systems theory. One of the first numerical simulations in biology was published in 1952 by the British neurophysiologists and Nobel prize winners Alan Lloyd Hodgkin and Andrew Fielding Huxley, who constructed a mathematical model that explained the action potential propagating along the axon of a neuronal cell. Their model described a cellular function emerging from the interaction between two different molecular components, a potassium and a sodium channel, and can therefore be seen as the beginning of computational systems biology. In 1960, Denis Noble developed the first computer model of the heart pacemaker. The formal study of systems biology, as a distinct discipline, was launched by systems theorist Mihajlo Mesarovic in 1966 with an international symposium at the Case Institute of Technology in Cleveland, Ohio entitled "Systems Theory and Biology".

The 1960s and 1970s saw the development of several approaches to study complex molecular systems, such as the Metabolic Control Analysis and the biochemical systems theory. The successes of molecular biology throughout the 1980s, coupled with a skepticism toward theoretical biology, that then promised more than it achieved, caused the quantitative modelling of biological processes to become a somewhat minor field. However the birth of functional genomics in the 1990s meant that large quantities of high quality data became available, while the computing power exploded, making more realistic models possible. In 1997, the group of Masaru Tomita published the first quantitative model of the metabolism of a whole (hypothetical) cell. Around the year 2000, after Institutes of Systems Biology were established in Seattle and Tokyo, systems biology emerged as a movement in its own right, spurred on by the completion of various genome projects, the large increase in data from the omics (e.g. genomics and proteomics) and the accompanying advances in high-throughput experiments and bioinformatics.

Since then, various research institutes dedicated to systems biology have been developed. For example, the NIGMS of NIH established a project grant that is currently supporting over ten Systems Biology Centers in the United States. As of summer 2006, due to a shortage of people in systems biology several doctoral training programs in systems biology have been established in many parts of the world. In that same year, the National Science Foundation (NSF) put forward a grand challenge for systems biology in the 21st century to build a mathematical model of the whole cell. In 2011, V. A. Shiva Ayyadurai and C. Forbes Dewey, Jr. of Department of Biological Engineering at the Massachusetts Institute of Technology created CytoSolve, a method to model the whole cell by dynamically integrating multiple molecular pathway models.

Associated Disciplines

According to the interpretation of Systems Biology as the ability to obtain, integrate and analyze complex data sets from multiple experimental sources using interdisciplinary tools, some typical technology platforms are:

- Phenomics: Organismal variation in phenotype as it changes during its life span.
- Genomics: Organismal deoxyribonucleic acid (DNA) sequence, including intra-organisamal cell specific variation. (i.e. Telomere length variation etc.).
- Epigenomics / Epigenetics: Organismal and corresponding cell specific transcriptomic regulating factors not empirically coded in the genomic sequence. (i.e. DNA methylation, Histone Acetelation etc.).
- Transcriptomics: Organismal, tissue or whole cell gene expression measurements by DNA microarrays or serial analysis of gene expression
- Interferomics: Organismal, tissue, or cell level transcript correcting factors (i.e. RNA interference)
- Translatomics / Proteomics: Organismal, tissue, or cell level measurements of proteins and peptides via two-dimensional gel electrophoresis, mass spectrometry or multi-dimensional protein identification techniques (advanced HPLC systems coupled with mass spectrometry). Sub disciplines include phosphoproteomics, glycoproteomics and other methods to detect chemically modified proteins.
- Metabolomics: Organismal, tissue, or cell level measurements of all small-molecules known as metabolites.
- Glycomics: Organismal, tissue, or cell level measurements of carbohydrates.
- Lipidomics: Organismal, tissue, or cell level measurements of lipids.

In addition to the identification and quantification of the above given molecules further techniques analyze the dynamics and interactions within a cell. This includes:

- Interactomics: Organismal, tissue, or cell level study of interactions between molecules. Currently the authoritative molecular discipline in this field of study is protein-protein interactions (PPI), although the working definition does not pre-clude inclusion of other molecular disciplines such as those defined here.

- NeuroElectroDynamics: Organismal, brain computing function as a dynamic system, underlying biophysical mechanisms and emerging computation by electrical interactions.
- Fluxomics: Organismal, tissue, or cell level measurements of molecular dynamic changes over time.
- Biomics: systems analysis of the biome.

The investigations are frequently combined with large-scale perturbation methods, including gene-based (RNAi, mis-expression of wild type and mutant genes) and chemical approaches using small molecule libraries. Robots and automated sensors enable such large-scale experimentation and data acquisition. These technologies are still emerging and many face problems that the larger the quantity of data produced, the lower the quality. A wide variety of quantitative scientists (computational biologists, statisticians, mathematicians, computer scientists, engineers, and physicists) are working to improve the quality of these approaches and to create, refine, and retest the models to accurately reflect observations.

The systems biology approach often involves the development of mechanistic models, such as the reconstruction of dynamic systems from the quantitative properties of their elementary building blocks. For instance, a cellular network can be modelled mathematically using methods coming from chemical kinetics and control theory. Due to the large number of parameters, variables and constraints in cellular networks, numerical and computational techniques are often used.

Other aspects of computer science and informatics are also used in systems biology. These include:

- New forms of computational model, such as the use of process calculi to model biological processes (notable approaches include stochastic ð-calculus, BioAmbients, Beta Binders, BioPEPA and Brane calculus) and constraint-based modeling.
- Integration of information from the literature, using techniques of information extraction and text mining.
- Development of online databases and repositories for sharing data and models, approaches to database integration and software interoperability via loose coupling of software, websites and databases, or commercial suits.
- Development of syntactically and semantically sound ways of representing biological models.

Up-regulation and Down-regulation

Up-regulation is a process that occurs within a cell triggered by a signal (originating internal or external to the cell), which results in increased expression of one or more genes and as a result the protein(s) encoded by those genes. On the converse, down-regulation is a process resulting in decreased gene and corresponding protein expression.

- Up-regulation occurs, for example, when a cell is deficient in some kind of receptor. In this case, more receptor protein is synthesized and transported to the membrane of the cell and, thus, the sensitivity of the cell is brought back to normal, reestablishing homeostasis.
- Down-regulation occurs, for example, when a cell is overstimulated by a neurotransmitter, hormone, or drug for a prolonged period of time, and the expression of the receptor protein is decreased in order to protect the cell .

Inducible vs. Repressible Systems

Gene Regulation can be summarized by the response of the respective system:

- Inducible systems - An inducible system is off unless there is the presence of some molecule (called an inducer) that allows for gene expression. The molecule is said to "induce expression". The manner by which this happens is dependent on the control mechanisms as well as differences between prokaryotic and eukaryotic cells.
- Repressible systems - A repressible system is on except in the presence of some molecule (called a corepressor) that suppresses gene expression. The molecule is said to "repress expression". The manner by which this happens is dependent on the control mechanisms as well as differences between prokaryotic and eukaryotic cells.

Theoretical Circuits

- Repressor/Inducer: an activation of a sensor results in the change of expression of a gene
- negative feedback: the gene product downregulates its own production directly or indirectly, which can result in
  - o keeping transcript levels constant/proportional to a factor
  - o inhibition of run-away reactions when coupled with a positive feedback loop
  - o creating an oscillator by taking advantage in the time delay of transcription and translation, given that the mRNA and protein half-life is shorter

- positive feedback: the gene product upregulates its own production directly or indirectly, which can result in
  - o signal amplification
  - o bistable switches when two genes inhibit each other and both have positive feedback
  - o pattern generation

Methods

In general, most experiments investigating differential expression used whole cell extracts of RNA, called steady-state levels, to determine which genes changed and by how much they did. These are, however, not informative of where the regulation has occurred and may actually mask conflicting regulatory processess but it is still the most commonly analysed (QPCR and DNA microarray).

When studying gene expression, there are several methods to look at the various stages. In eukaryotes these include:

- The local chromatin environment of the region can be determined by ChIP-chip analysis by pulling down RNA Polymerase II, Histone 3 modifications, Trithorax-group protein, Polycomb-group protein, or any other DNA-binding element to which a good antibody is available.
- Epistatic interactions can be investigated by synthetic genetic array analysis
- Due to post-transcriptional regulation, transcription rates and total RNA levels differ significantly. To measure the transcription rates nuclear run-on assays can be done and newer high-throughput methods are being developed, using thiol labelling instead of radioactivity.
- Only 5% of the RNA polymerised in the nucleus actually exists, and not only introns, abortive products, and non-sense transcripts are degradated. Therefore, the differences in nuclear and cytoplasmic levels can be see by separating the two fractions by gentle lysis.
- Alternative splicing can be analysed with a splicing array or with a tiling array.
- All in vivo RNA is complexed as RNPs. The quantity of transcripts bound to specific protein can be also analysed by RIP-Chip. For example, DCP2 will give an indication of sequestered protein; ribosome-bound gives and indication of transcripts active in transcription (although it should be noted that a more dated method, called polysome fractionation, is still popular in some labs)
- Protein levels can be analysed by Mass spectrometry, which can be compared only to QPCR data, as microarray data is relative and not absolute.

- RNA and protein degradation rates are measured by means of transcription inhibitors (actinomycin D or á-amanitin) or translation inhibitors (Cycloheximide), respectively.

Nucleic Acid Methods

Nucleic acid methods are the techniques used to study nucleic acids (DNA and RNA).

Purification

- Phenol-chloroform extraction
- minicolumn purification
- RNA extraction

Quantification

- Abundance in weight: spectroscopic quantification
- Absolute abundance in number: Q-PCR
- high-throughput relative abundance DNA microarray
- high-throughput absolute abundance SAGE
- Size: Gel electrophoresis

Synthesis

- De novo: Oligonucleotide synthesis
- Amplification: PCR

Kinetics

- Dual polarization interferometry
- QCM-D

Other

- Nucleic acid simulations
- DNA sequencing
- Bisulfite sequencing
- Expression cloning
- Northern blot
- Sucrose gradient centrifugation
- Radioactivity in biological research
- Lab-on-a-chip
- Nuclear run-on assay
- Fluorescent in situ hybridization
- several Bioinformatics methods, such as RNA structure prediction

Protein Methods

Protein methods are the techniques used to study proteins.

There are genetic methods for studying proteins, methods for detecting proteins, methods for isolating and purifying proteins and other methods for characterizing the structure and function of proteins, often requiring that the protein first be purified.

Genetic Methods

- conceptual translation- many proteins are never directly sequenced, but their sequence of amino acids is known by "conceptual translation" of a known mRNA sequence. See genetic code.
- site-directed mutagenesis allows new variants of proteins to be produced and tested for how structural changes alter protein function.
  - o insertion of protein tags such as the His-tag. See also green fluorescent protein.
- evolutionary; analysis of sequence changes in different species using software such as BLAST.
- Proteins that are involved in human diseases can be identified by matching alleles to disease and other phenotypes using methods such as calculation of LOD scores.

Detecting Proteins

- microscopy, protein immunostaining
- Protein immunoprecipitation
- Immunoelectrophoresis
- Immunoblotting
- BCA Protein Assay
- Western blot
- Spectrophotometry
- Enzyme assay

Protein Purification

- Protein Isolation
  - o chromatography methods: Ion exchange, Size-exclusion chromatography or Gelfiltration, Affinity chromatography
- Protein Extraction and Solubilization
- Protein Concentration Determination Methods, Bradford protein assay
- Concentrating Protein Solutions

- Gel electrophoresis
  - o Gel Electrophoresis Under denaturing conditions
  - o Gel Electrophoresis Under non-denaturing conditions
  - o 2D Gel Electrophoresis
- Electrofocusing

Protein Structures

- X-ray crystallography
- Protein NMR

Protein-DNA Interactions

- ChIP-on-chip
- Chip-Sequencing
- DamID
- Microscale Thermophoresis

Other Methods

- Hydrogen-deuterium exchange
- Mass spectrometry
- Molecular dynamics
- Protein structure prediction
- Protein sequencing
- Protein structural alignment
- Protein ontology
- Protein synthesis
- Proteomics
- Peptide mass fingerprinting
- Ligand binding assay
- Eastern blotting
- metabolic labeling
  - o heavy isotope labeling
  - o radioactive isotope labeling
- Enhancer (genetics)

In genetics, an enhancer is a short region of DNA that can be bound with proteins (namely, the trans-acting factors, much like a set of transcription factors) to enhance transcription levels of genes (hence the name) in a gene cluster. While enhancers are usually cis-acting, an enhancer does not need to be particularly close to the genes it acts on, and sometimes need not be located on the same chromosome. Enhancer locations

In eukaryotic cells the structure of the chromatin complex of DNA is folded in a way that functionally mimics the supercoiled state characteristic of prokaryotic DNA, so that although the enhancer DNA is far from the gene in regard to the number of nucleotides, it is geometrically close to the promoter and gene. This allows it to interact with the general transcription factors and RNA polymerase II.

An enhancer may be located upstream or downstream of the gene it regulates. Furthermore, an enhancer need not be located near to the transcription initiation site to affect transcription, as some have been found located in several hundred thousand base pairs upstream or downstream of the start site. Enhancers do not act on the promoter region itself, but are bound by activator proteins.

These activator proteins interact with the mediator complex, which recruits polymerase II and the general transcription factors which then begin transcribing the genes. Enhancers can also be found within introns. An enhancer's orientation may even be reversed without affecting its function. Additionally, an enhancer may be excised and inserted elsewhere in the chromosome, and still affect gene transcription. That is one reason that intron polymorphisms may have effects although they are not translated.

Theories

Currently, there are two different theories on the information processing that occurs on enhancers:

- Enhanceosomes - rely on highly cooperative, coordinated action and can be disabled by single point mutations that move or remove the binding sites of individual proteins.
- Flexible billboards - less integrative, multiple proteins independently regulate gene expression and their sum is read in by the basal transcriptional machinery

HACNS1

HACNS1 (also known as CENTG2 and located in the Human Accelerated Region 2) is a gene enhancer "that may have contributed to the evolution of the uniquely opposable human thumb, and possibly also modifications in the ankle or foot that allow humans to walk on two legs". Evidence to date shows that of the 110,000 gene enhancer sequences identified in the human genome, HACNS1 has undergone the most change during the evolution of humans following the split with the ancestors of chimpanzees.

GADD45G

An enhancer near the gene GADD45g has been described that may regulate brain growth in chimpanzees and other mammals, but not in humans. The GADD45G regulator in mice and chimps is active in regions of the brain where cells that form the cortex, ventral forebrain, and thalamus are located and may suppress further neurogenesis. Loss of the GADD45G enhancer in humans may contribute to an increase of certain neuronal populations and to forebrain expansion in humans.

Enhancers in Developmental Biology

The development, differentiation and growth of cells and tissues require precisely regulated patterns of gene expression. Enhancers work as cis-regulatory elements to mediate both spatial and temporal control of development by turning on transcription in specific cells and/or repressing it in other cells. Thus, the particular combination of transcription factors and other DNA-binding proteins in a developing tissue controls which genes will be expressed in that tissue. Enhancers allow the same gene to be used in diverse processes in space and time.

Identification and Characterization

Enhancers can be identified by enhancer trap techniques using a reporter gene or by comparative sequence analysis and computational genomics. In genetically tractable models such as the fruit fly Drosophila melanogaster, for example, a reporter construct such as the lacZ gene can be randomly integrated into the genome using a P element transposon. If the reporter gene integrates near an enhancer, its expression will reflect the expression pattern driven by that enhancer. Thus, staining the flies for LacZ expression or activity and cloning the sequence surrounding the integration site can identify the enhancer sequence.

In the comparative genomics approach, sequence conservation of non-coding regions can be indicative of enhancers. Sequences from multiple species are aligned, and conserved regions are identified computationally. Identified sequences can then be attached to a reporter gene such green fluorescent protein or lacZ to determine the in vivo pattern of gene expression produced by the enhancer when injected into an embryo. mRNA expression of the reporter can be visualized by in situ hybridization, which provides a more direct measure of enhancer activity, since it is not subjected to the complexities of translation and protein folding. Although much evidence has pointed to sequence conservation for critical developmental enhancers, other work has shown that the function of enhancers can be conserved with little or no primary sequence conservation. For example, the RET enhancers in humans have very little sequence conservation to those in zebrafish, yet both species' sequences produce nearly identical patterns of reporter gene expression in zebrafish. Additionally, algorithms such as TFSEARCH or JASPAR can be used to identify putative transcription factor binding sites to further characterize an enhancer sequence.

Enhancers in Invertebrate Segmentation

The enhancers determining early segmentation in Drosophila melanogaster embryos are among the best characterized developmental enhancers. In the early fly embryo, the gap gene transcription factors are responsible for activating and repressing a number of segmentation genes, such as the pair rule genes. The gap genes are expressed in blocks along the anterior-posterior axis of the fly along with other maternal effect transcription factors, thus creating zones within which different combinations of transcription factors are expressed. The pair-rule genes are separated from one another by non-expressing cells. Moreover, the stripes of expression for different pair-rule genes are offset by a few cell diameters from one another. Thus, unique combinations of pair-rule gene expression create spatial domains along the anterior-posterior axis to set up each of the 14 individual segments. The 480 bp enhancer responsible for driving the sharp stripe two of the pair-rule gene even-skipped (eve) has been well-characterized. The enhancer contains 12 different binding sites for maternal and gap gene transcription factors. Activating and repressing sites overlap in sequence. Eve is only expressed in a narrow stripe of cells that contain high concentrations of the activators and low concentration of the repressors for this enhancer sequence. Other enhancer regions drive eve expression in 6 other stripes in the embryo.

Enhancers in Vertebrate Patterning

Establishing body axes is a critical step in animal development. During mouse embryonic development, Nodal, a transforming growth factor-beta superfamily ligand, is a key gene involved in patterning both the anterior-posterior axis and the left-right axis of the early embryo. The Nodal gene contains two enhancers: the Proximal Epiblast Enhancer (PEE) and the Asymmetric Enhancer (ASE). The PEE is upstream of the Nodal gene and drives Nodal expression in the portion of the primitive streak that will differentiate into the node (also referred to as the primitive node). The PEE turns on Nodal expression in response to a combination of Wnt signaling plus a second, unknown signal; thus, a member of the LEF/TCF transcription factor family likely binds to a TCF binding site in the cells in the node. Diffusion of Nodal away from the node forms a gradient which then patterns the extending anterior-posterior axis of the embryo. The ASE is an intronic enhancer bound by the fork head domain transcription factor Fox1. Early in development,

Fox1-driven Nodal expression establishes the visceral endoderm. Later in development, Fox1 binding to the ASE drives Nodal expression on the left side of the lateral plate mesoderm, thus establishing left-right asymmetry necessary for asymmetric organ development in the mesoderm.

Establishing three germ layers during gastrulation is another critical step in animal development. Each of the three germ layers has unique patterns of gene expression that promote their differentiation and development. The endoderm is specified early in development by Gata4 expression, and Gata4 goes on to direct gut morphogenesis later. Gata4 expression is controlled in the early embryo by an intronic enhancer that binds another forkhead domain transcription factor, FoxA2. Initially the enhancer drives broad gene expression throughout the embryo, but the expression quickly becomes restricted to the endoderm, suggesting that other repressors may be involved in its restriction. Late in development, the same enhancer restricts expression to the tissues that will become the stomach and pancreas. An additional enhancer is responsible for maintaining Gata4 expression in the endoderm during the intermediate stages of gut development.

Multiple Enhancers Promote Developmental Robustness

Some genes involved in critical developmental processes contain multiple enhancers of overlapping function. Secondary enhancers, or "shadow enhancers," may be found many kilobases away from the primary enhancer ("primary" usually refers to the first enhancer discovered, which is often closer to the gene it regulates). On its own, each enhancer drives nearly identical patterns of gene expression. Are the two enhancers truly redundant? Recent work has shown that multiple enhancers allow fruit flies to survive environmental perturbations, such as an increase in temperature. When raised at an elevated temperature, a single enhancer sometimes fails to drive the complete pattern of expression, whereas the presence of both enhancers permits normal gene expression.

Enhancers and the Evolution of Developmental Mechanisms

One theme of research in evolutionary developmental biology ("evo-devo") is investigating the role of enhancers and other cis-regulatory elements in producing morphological changes via developmental differences between species.

Stickleback PitX1

Recent work has investigated the role of enhancers in morphological changes in threespine stickleback fish. Sticklebacks exist in both marine and freshwater environments, but sticklebacks in many freshwater populations have completely lost their pelvic fins (appendages homologous to the posterior limb of tetrapods). Pitx1 is a homeobox gene involved in posterior limb development in vertebrates. Preliminary genetic analyses indicated that changes in the expression of this gene were responsible for pelvic reduction in sticklebacks. Fish expressing only the freshwater allele of Pitx1 do not have pelvic spines, whereas fish expressing a marine allele retain pelvic spines.

A more thorough characterization showed that a 500 base pair enhancer sequence is responsible for turning on Pitx1 expression in the posterior fin bud. This enhancer is located near a chromosomal fragile site—a sequence of DNA that is likely to be broken and thus more likely to be mutated as a result of imprecise DNA repair. This fragile site has caused repeated, independent losses of the enhancer responsible for driving PitX1 expression in the pelvic spines in isolated freshwater population, and without this enhancer, freshwater fish fail to develop pelvic spines.

Enhancers in Drosophila wing Pattern Evolution

Pigmentation patterns provide one of the most striking and easily scored differences between different species of animals. Pigmentation of the Drosophila wing has proven to be a particularly amenable system for studying the development of complex pigmentation phenotypes. The Drosophila guttifera wing has 12 dark pigmentation spots and 4 lighter gray intervein patches. Pigment spots arise from expression of the yellow gene, whose product produces black melanin.

Recent work has shown that two enhancers in the yellow gene produce gene expression in precisely this pattern – the vein spot enhancer drives reporter gene expression in the 12 spots, and the intervein shade enhancer drives reporter expression in the 4 distinct patches. These two enhancers are responsive to the Wnt signaling pathway, which is activated by "wingless" expression at all of the pigmented locations. Thus, in the evolution of the complex pigmentation phenotype, the yellow pigment gene evolved enhancers responsive to the wingless signal and wingless expression evolved at new locations to produce novel wing patterns.

- Artificial transcription factors (small molecules that mimic transcription factor protein)

An artificial transcription factor is an example of a chimeral protein, designed to target and modulate gene transcription. They are generally composed of a DNA-binding domain (specific to a certain sequence) coupled to a modulatory domain (which acts upon other transcription factors) in order to alter the expression of a particular gene. It is also possible to downregulate expression of a gene by targeting the 5' untranslated region with a DNA-binding domain that lacks a regulatory domain; this will reduce transcription simply by blocking RNA polymerase progression along the DNA template.

Cellular model

Creating a cellular model has been a particularly challenging task of systems biology and mathematical biology. It involves developing efficient algorithms, data structures, visualization and communication tools to orchestrate the integration of large quantities of biological data with the goal of computer modeling.

It is also directly associated with bioinformatics, computational biology and Artificial life. It involves the use of computer simulations of the many cellular subsystems such as the networks of metabolites and enzymes which comprise metabolism, signal transduction pathways and gene regulatory networks to both analyze and visualize the complex connections of these cellular processes.

The complex network of biochemical reaction/transport processes and their spatial organization make the development of a predictive model of a living cell a grand challenge for the 21st century.

Overview

The eukaryotic cell cycle is very complex and is one of the most studied topics, since its misregulation leads to cancers. It is possibly a good example of a mathematical model as it deals with simple calculus but gives valid results.

Two research groups have produced several models of the cell cycle simulating several organisms. They have recently produced a generic eukaryotic cell cycle model which can represent a particular eukaryote depending on the values of the parameters, demonstrating that the idiosyncrasies of the individual cell cycles are due to different protein concentrations and affinities, while the underlying mechanisms are conserved (Csikasz-Nagy et al., 2006).

By means of a system of ordinary differential equations these models show the change in time (dynamical system) of the protein inside a single typical cell; this type of model is called a deterministic process (whereas a model describing a statistical distribution of protein concentrations in a population of cells is called a stochastic process).

To obtain these equations an iterative series of steps must be done: first the several models and observations are combined to form a consensus diagram and the appropriate kinetic laws are chosen to write the differential equations, such as rate kinetics for stoichiometric reactions, Michaelis-Menten kinetics for enzyme substrate reactions and Goldbeter–Koshland kinetics for ultrasensitive transcription factors, afterwards the parameters of the equations (rate constants, enzyme efficiency coefficients and Michealis constants) must be fitted to match observations; when they cannot be fitted the kinetic equation is revised and when that is not possible the wiring diagram is modified. The parameters are fitted and validated using observations of both wild type and mutants, such as protein half-life and cell size.

In order to fit the parameters the differential equations need to be studied. This can be done either by simulation or by analysis.

In a simulation, given a starting vector (list of the values of the variables), the progression of the system is calculated by solving the equations at each time-frame in small increments.

In analysis, the proprieties of the equations are used to investigate the behavior of the system depending of the values of the parameters and variables.

A system of differential equations can be represented as a vector field, where each vector described the change (in concentration of two or more protein) determining where and how fast the trajectory (simulation) is heading.

Vector fields can have several special points: a stable point, called a sink, that attracts in all directions (forcing the concentrations to be at a certain value), an unstable point, either a source or a saddle point which repels (forcing the concentrations to change away from a certain value), and a limit cycle, a closed trajectory towards which several trajectories spiral towards (making the concentrations oscillate).

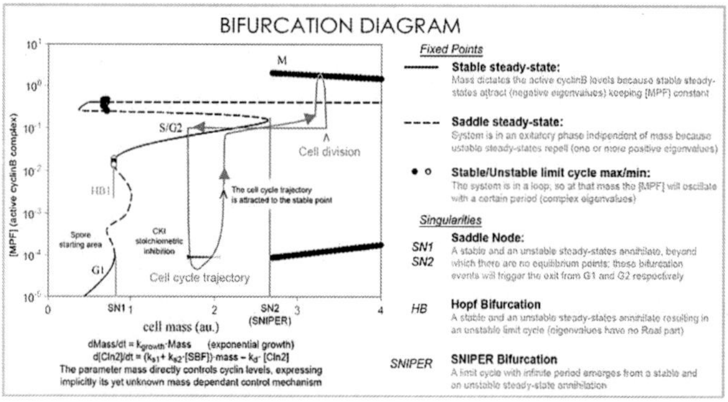

A better representation which can handle the large number of variables and parameters is called a bifurcation diagram (bifurcation theory): the presence of these special steady-state points at certain values of a parameter (e.g. mass) is represented by a point and once the parameter passes a certain value, a qualitative change occurs, called a bifurcation, in which the nature of the space changes, with profound consequences for the protein concentrations: the cell cycle has phases (partially corresponding to G1 and G2) in which mass, via a stable point, controls cyclin levels, and phases (S and M phases) in which the concentrations change independently, but once the phase has changed at a bifurcation event (cell cycle checkpoint), the system cannot go back to the previous levels since at the current mass the vector field is profoundly different and the mass cannot be reversed back through the bifurcation event, making a checkpoint irreversible.

In particular the S and M checkpoints are regulated by means of special bifurcations called a Hopf bifurcation and an infinite period bifurcation.

Molecular Level Simulations

E-Cell Project aims "to make precise whole cell simulation at the molecular level possible".

Projects

Multiple projects are in progress.

- Karyote - Indiana University
- E-Cell Project
- Virtual Cell - University of Connecticut Health Center
- Silicon Cell

Biological Data Visualization

Biology Data Visualization is a branch of bioinformatics concerned with the application of computer graphics, scientific visualization, and information visualization to different areas of the life sciences. This includes visualization of sequences, genomes, alignments, phylogenies, macromolecular structures, systems biology, microscopy, and magnetic resonance imaging data. Software tools used for visualizing biological data range from simple, stand-alone programs to complex, integrated systems.

State-of-the-art and Perspectives

Today we are experiencing a rapid growth in volume and diversity of biological data, presenting an increasing challenge for biologists. A key step in understanding and learning from these data is visualization. Thus, there has been a corresponding increase in the number and diversity of systems for visualizing biological data.

An emerging trend is the blurring of boundaries between the visualization of 3D structures at atomic resolution, visualization of larger complexes by cryo-electron microscopy, and visualization of the location of proteins and complexes within whole cells and tissues.

A second emerging trend is an increase in the availability and importance of time-resolved data from systems biology, electron microscopy and cell and tissue imaging. In contrast, visualization of trajectories has long been a prominent part of molecular dynamics.

Finally, as datasets are increasing in size, complexity, and interconnectness, biological visualization systems are improving in usability, data integration and standardization.

List of Visualization Systems

A large number of software systems are available for visualization biological data. The links below link lists of such systems, grouped by application areas.

Conserved Non-coding DNA sequence

A conserved non-coding sequence (CNS) is a DNA sequence of noncoding DNA that is evolutionarily conserved. These sequences are of interest for their potential to regulate gene production.

CNSs in plants and animals are highly associated with transcription factor binding sites and other cis-acting regulatory elements. Conserved non-coding sequences can be important sites of evolutionary divergence as mutations in these regions may alter the regulation of conserved genes, producing species-specific patterns of gene expression.

Spatiotemporal Gene Expression

Spatiotemporal gene expression is the activation of genes within specific tissues of an organism at specific times during development. Gene activation patterns vary widely in complexity. Some are straightforward and static, such as the pattern of tubulin, which is expressed in all cells at all times in life. Some, on the other hand, are extraordinarily intricate and difficult to predict and model, with expression fluctuating wildly from minute to minute or from cell to cell. Spatiotemporal variation plays a key role in generating the diversity of cell types found in developed organisms; since the identity of a cell is specified by the collection of genes actively expressed within that cell, if gene expression was uniform spatially and temporally, there could be at most one kind of cell.

Consider the gene wingless, a member of the wnt family of genes. In the early embryonic development of the model organism Drosophila melanogaster, or fruit fly, wingless is expressed across almost the entire embryo in alternating stripes three cells separated. This pattern is lost by the time the organism develops into a larva, but wingless is still expressed in a variety of tissues such as the wing imaginal discs, patches of tissue that will develop into the adult wings. The spatiotemporal pattern of wingless gene expression is determined by a network of regulatory interactions consisting of the effects of many different genes such as even-skipped and Krüppel.

What causes spatial and temporal differences in the expression of a single gene? Because current expression patterns depend strictly on previous expression patterns, there is a regressive problem of explaining what caused the first differences in gene expression. The process by which uniform gene expression becomes spatially and temporally differential is known as symmetry breaking. For example, in the case of embryonic Drosophila development, the genes nanos and bicoid are asymmetrically expressed in the oocyte because maternal cells deposit messenger RNA (mRNA) for these genes in the poles of the egg before it is laid.

Identifying Spatiotemporal Patterns

One way to identify the expression pattern of a particular gene is to place a reporter gene downstream of its promoter. In this configuration, the promoter gene will cause the reporter gene to be expressed only where and when the gene of interest is expressed. The expression distribution of the reporter gene can be determined by visualizing it. For example, the reporter gene green fluorescent protein can be visualized by stimulating it with blue light and then using a digital camera to record green fluorescent emission.

If the promoter of the gene of interest is unknown, there are several ways to identify its spatiotemporal distribution. Immunohistochemistry involves preparing an antibody with specific affinity for the protein associated with the gene of interest. This distribution of this antibody can then be visualized by a technique such as fluorescent labeling. Immunohistochemistry has the advantages of being methodologically feasible and relatively inexpensive. Its disadvantages include non-specificity of the antibody leading to false positive identification of expression. Poor penetrance of the antibody into the target tissue can lead to false negative results. Furthermore, since immunohistochemistry visualizes the protein generated by the gene, if the protein product diffuses between cells, or has a particularly short or long half-life relative to the mRNA that is used to translate the protein, this can lead to distorted interpretation of which cells are expressing the mRNA.

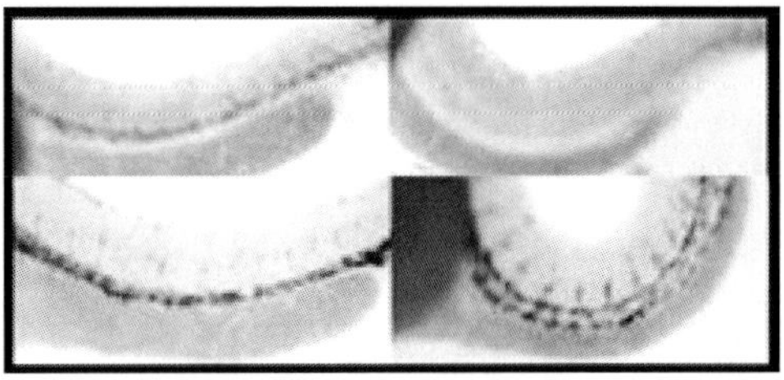

In situ hybridizations of genes expressed in arteries (top) and veins (bottom) in zebrafish. Blue staining indicates presence of the gene mRNAs. Panels on the left are normal animals, while animals on the right are mutated in the Notch gene. Fish lacking Notch have fewer arteries and more veins at this point in developmental time.

In situ hybridization is an alternate method in which a "probe," a synthetic nucleic acid with a sequence complementary to the mRNA of the gene, is added to the tissue. This probe is then chemically tagged so that it can be visualized later. This technique enables visualization specifically of mRNA-producing cells without any of the artifacts associated with immunohistochemistry. However, it is notoriously difficult, and requires knowledge of the sequence of DNA corresponding to the gene of interest.

A method called enhancer-trap screening reveals the diversity of spatiotemporal gene expression patterns possible in an organism. In this technique, DNA that encodes a reporter gene is randomly inserted into the genome. Depending on the gene promoters proximal to the insertion point, the reporter gene will be expressed in particular tissues at particular points in development. While enhancer-trap derived expression patterns do not necessarily reflect the actual patterns of expression of specific genes, they reveal the variety of spatiotemporal patterns that are accessible to evolution.

Reporter genes can be visualized in living organisms, but both immunohistochemistry and in situ hybridization must be performed in fixed tissues. Techniques that require fixation of tissue can only generate a single temporal time point per individual organism. However, using live animals instead of fixed tissue can be crucial in dynamically understanding expression patterns over an individual's lifespan. Either way, variation between individuals can confound the interpretation of temporal expression patterns.

Post-transcriptional Regulation

In eukaryotes, where export of RNA is required before translation is possible, nuclear export is thought to provide additional control over gene expression. All transport in and out of the nucleus is via the nuclear pore and transport is controlled by a wide range of importin and exportin proteins.

Expression of a gene coding for a protein is only possible if the messenger RNA carrying the code survives long enough to be translated. In a typical cell an RNA molecule is only stable if specifically protected from degradation. RNA degradation has particular importance in regulation of expression in eukaryotic cells where mRNA has to travel significant distances before being translated. In eukaryotes RNA is stabilised by certain post-transcriptional modifications, particularly the 5' cap and poly-adenylated tail.

Intentional degradation of mRNA is used not just as a defence mechanism from foreign RNA (normally from viruses) but also as a route of mRNA destabilisation. If an mRNA molecule has a complementary sequence to a small interfering RNA then it is targeted for destruction via the RNA interference pathway.

Translational Regulation

Neomycin is an example of a small molecule which reduces expression of all protein genes inevitably leading to cell death, thus acts as an antibiotic.

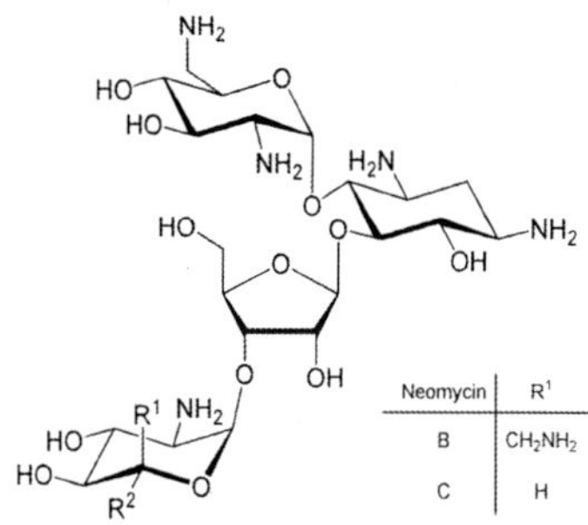

Neomycin	R^1	R^2
B	CH_2NH_2	H
C	H	CH_2NH_2

Direct regulation of translation is less prevalent than control of transcription or mRNA stability but is occasionally used. Inhibition of protein translation is a major target for toxins and antibiotics in order to kill a cell by overriding its normal gene expression control. Protein synthesis inhibitors include the antibiotic neomycin and the toxin ricin.

Protein Degradation

Once protein synthesis is complete the level of expression of that protein can be reduced by protein degradation. There are major protein degradation pathways in all prokaryotes and eukaryotes of which the proteasome is a common component. An unneeded or damaged protein is often labelled for degradation by addition of ubiquitin.

Measurement

Measuring gene expression is an important part of many life sciences - the ability to quantify the level at which a particular gene is expressed within a cell, tissue or organism can give a huge amount of information. For example measuring gene expression can:

- Identify viral infection of a cell (viral protein expression)
- Determine an individual's susceptibility to cancer (oncogene expression)
- Find if a bacterium is resistant to penicillin (beta-lactamase expression)

Similarly the analysis of the location of expression protein is a powerful tool and this can be done on an organism or cellular scale. Investigation of localisation is particularly important for study of development in multicellular organisms and as an indicator of protein function in single cells. Ideally measurement of expression is done by detecting the final gene product (for many genes this is the protein) however it is often easier to detect one of the precursors, typically mRNA, and infer gene expression level.

mRNA Quantification

Levels of mRNA can be quantitatively measured by northern blotting which gives size and sequence information about the mRNA molecules. A sample of RNA is separated on an agarose gel and hybridized to a radio-labeled RNA probe that is complementary to the target sequence.

The radio-labeled RNA is then detected by an autoradiograph. The main problems with Northern blotting stem from the use of radioactive reagents (which make the procedure time consuming and potentially dangerous) and lower quality quantification than more modern methods (due to the fact that quantification is done by measuring band strength in an image of a gel). Northern blotting is, however, still widely used as the additional mRNA size information allows the discrimination of alternately spliced transcripts.

A more modern low-throughput approach for measuring mRNA abundance is reverse transcription quantitative polymerase chain reaction (RT-PCR followed with qPCR). RT-PCR first generates a DNA template from the mRNA by reverse transcription, which is called cDNA. This cDNA template is then used for qPCR where the change in fluorescence of a probe changes as the DNA amplification process progresses. With a carefully constructed standard curve qPCR can produce an absolute measmsensitive (detection of a single mRNA molecule is possible), but can be expensive due to the fluorescent probes required.

An even more advanced approach is to individually tag single mRNA molecules with fluorescent barcodes (nanostrings), which can be detected one-by-one and counted for direct digital quantification. The advantage of this approach is that it does not rely on analog quantification of fluorescent intensity, which can be problematic due to noise, lack of linearity, and narrow dynamic range. Instead, the technique relies on fluorescence to detect simply the presence of a single mRNA molecule in a binary ("yes" or "no") mode. This method was invented by Dr. Krassen Dimitrov at the Institute for Systems Biology and commercialized through his start-up company, NanoString Technologies

Northern blots and RT-qPCR are good for detecting whether a single gene is being expressed, but it quickly becomes impractical if many genes within the sample are being studied. Using DNA microarrays, transcript levels for many genes at once (expression profiling) can be measured. Recent advances in microarray technology allow for the quantification, on a single array, of transcript levels for every known gene in several organism's genomes, including humans. Alternatively "tag based" technologies like Serial analysis of gene expression (SAGE), which can provide a relative measure of the cellular concentration of different mRNAs, can be used. The great advantage of tag-based methods is the "open architecture", allowing for the exact measurement of any transcript, with a known or unknown sequence.

Protein Quantification

For genes encoding proteins the expression level can be directly assessed by a number of means with some clear analogies to the techniques for mRNA quantification.

The most commonly used method is to perform a Western blot against the protein of interest - this gives information on the size of the protein in addition to its identity. A sample (often cellular lysate) is separated on a polyacrylamide gel, transferred to a membrane and then probed with an antibody to the protein of interest. The antibody can either be conjugated to a fluorophore or to horseradish peroxidase for imaging and/or quantification. The gel-based nature of this assay makes quantification less accurate but it has the advantage of being able to identify later modifications to the protein, for example proteolysis or ubiquitination, from changes in size.

Localisation

In situ-hybridization of Drosophila embryos at different developmental stages for the mRNA responsible for the expression of hunchback. High intensity of blue color marks places with high hunchback mRNA quantity. Analysis of expression is not limited to only quantification; localisation can also be determined. mRNA can be detected with a suitably labelled complementary mRNA strand and protein can be detected via labelled antibodies.

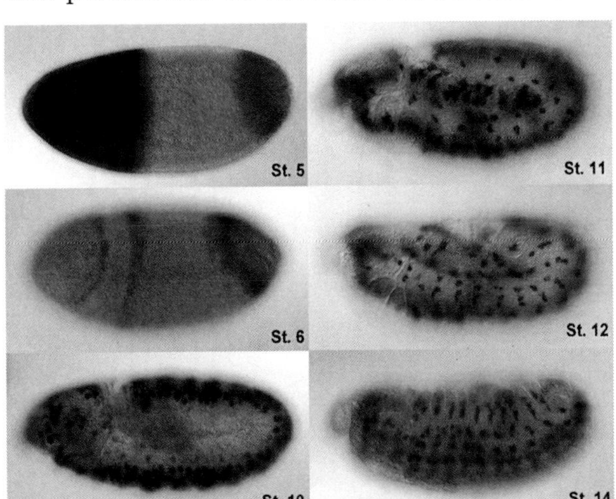

The probed sample is then observed by microscopy to identify where the mRNA or protein is. By replacing the gene with a new version fused a green fluorescent protein (or similar) marker expression may be directly quantified in live cells. This is done by imaging using a fluorescence microscope. It is very difficult to clone a GFP-fused protein into its native location in the genome without affecting expression levels so this method often cannot be used to measure endogenous gene expression. It is, however, widely used to measure the expression of a gene artificially introduced into the cell, for example via an expression vector. It is important to note that by fusing a target protein to a fluorescent reporter the protein's behavior, including its cellular localization and expression level, can be significantly changed.

The enzyme-linked immunosorbent assay works by using antibodies immobilised on a microtiter plate to capture proteins of interest from samples added to the well. Using a detection antibody conjugated to an enzyme or fluorophore the quantity of bound protein can be accurately measured by fluorometric or colourimetric detection. The detection process is very similar to that of a Western blot, but by avoiding the gel steps moreaccurate quantification can be achieved.

Expression System

An expression system is a system specifically designed for the production of a gene product of choice. This is normally a protein although may also be RNA, such as tRNA or a ribozyme. An expression system consists of a gene, normally encoded by DNA, and the molecular machinery required to transcribe the DNA into mRNA and translate the mRNA into protein using the reagents provided.

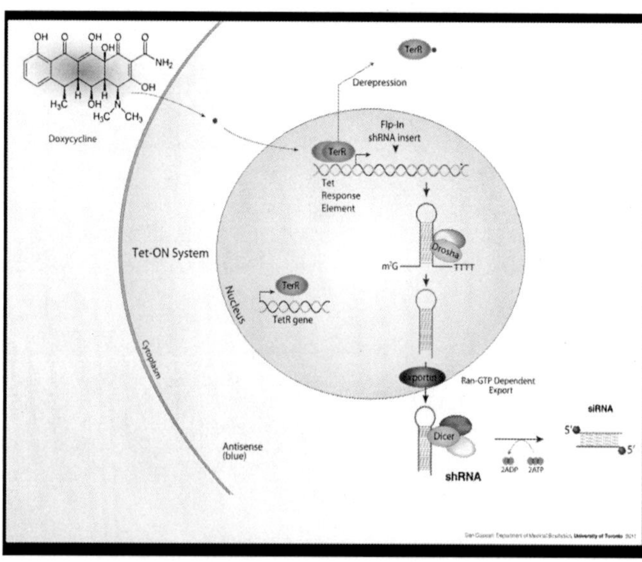

Figure: *Tet-ON inducible shRNA system*

In the broadest sense this includes every living cell but the term is more normally used to refer to expression as a laboratory tool. An expression system is therefore often artificial in some manner. Expression systems are, however, a fundamentally natural process. Viruses are an excellent example where they replicate by using the host cell as an expression system for the viral proteins and genome.

Inducible Expression

Doxycycline is also used in "Tet-on" and "Tet-off" tetracycline controlled transcriptional activation to regulate transgene expression in organisms and cell cultures

In Nature

In addition to these biological tools, certain naturally observed configurations of DNA (genes, promoters, enhancers, repressors) and the associated machinery itself are referred to as an expression system. This term is normally used in the case where a gene or set of genes is switched on under well defined conditions. For example the simple repressor switch expression system in Lambda phage and the lac operator system in bacteria. Several natural expression systems are directly used or modified and used for artificial expression systems such as the Tet-on and Tet-off expression system.

2

Molecular Mechanisms

The molecular basis of dominance was unknown to Mendel. It is now understood that a gene locus includes a long series (hundreds to thousands) of bases or nucleotides of deoxyribonucleic acid (DNA) at a particular point on a chromosome. The central dogma of molecular biology states that "DNA makes RNA makes protein", that is, that DNA is transcribed to make an RNA copy, and RNA is translated to make a protein. In this process, different alleles at a locus may or may not be transcribed, and if transcribed may be translated to slightly different forms of the same protein (called isoforms).

Proteins often function as enzymes that catalyze chemical reactions in the cell, which directly or indirectly produce phenotypes. In any diploid organism, the DNA sequences of the two alleles present at any gene locus may be identical (homozygous) or different (heterozygous). Even if the gene locus is heterozygous at the level of the DNA sequence, the proteins made by each allele may be identical.

In the absence of any difference between the protein products, neither allele can be said to be dominant (see co-dominance, above). Even if the two protein products are slightly different (allozymes), it is likely that they produce the same phenotype with respect to enzyme action, and again neither allele can be said to be dominant.

Dominance typically occurs when one of the two alleles is non-functional at the molecular level, that is, it is not transcribed or else does not produce a protein product. This can be the result of a mutation that alters the DNA sequence of the allele. An organism homozygous for the non-functional allele will generally show a distinctive phenotype, due to the absence of the protein product. For example, in humans and other organisms, the unpigmented skin of the albino phenotype results when an individual is homozygous for an allele that prevents synthesis of the skin pigment protein melanin.

It is important to understand that it is not the lack of function that allows the allele to be described as recessive: this is the interaction with the alternative allele in the heterozygote. Three general types of interaction are possible:

1. In the typical case, the single functional allele makes sufficient protein to produce a phenotype identical to that of the homozygote: this is called haplosufficiency. For example, suppose the standard amount of enzyme produced in the functional homozygote is 100%, with the two functional alleles contributing 50% each. The single functional allele in the heterozygote produces 50% of the standard amount of enzyme, which is sufficient to produce the standard phenotype. If the heterozygote and the functional-allele homozygote have identical phenotypes, the functional allele is dominant to the non-functional allele. This occurs at the albino gene locus: the heterozygote produces sufficient enzyme to convert the pigment precursor to melanin, and the individual has standard pigmentation.
2. Alternatively, a single functional allele in the heterozygote may produce insufficient gene product for proper function, and the phenotype resembles that of the homozygote for the non-functional allele. This haploinsufficiency is much less common: usually the deficiency of gene product results in incomplete dominance (below).
3. The intermediate interaction occurs where the heterozygous genotype produces a phenotype intermediate between the two homozygotes. Depending on which of the two homozygotes the heterozygote most resembles, one allele is said to show incomplete dominance over the other. For example, in humans the Hb gene locus is responsible for the Beta-chain protein (HBB) that is one of the two globin proteins that make up the blood pigment hemoglobin Many people are homozygous for an allele called Hb ; some persons carry an alternative allele called Hb., either as homozygotes or heterozygotes. The hemoglobin molecules of Hb /Hb homozygotes, undergo a change in shape that distorts the morphology of the red blood cells, and causes a severe, life-threatening form of anemia called sickle-cell anemia. Persons heterozygous Hb /Hb for this allele have a much less severe form of anemia called sickle-cell trait. Because the disease phenotype of Hb /Hb heterozygotes is more similar to but not identical to the Hb /Hb homozygote, the Hb allele is said to be incompletely dominant to the Hb allele.

In some cases, dominance of a non-standard allele results when that allele produces a defective protein that interferes with the proper function of the protein produced by the standard allele. The presence of the defective protein "dominates" the standard protein, and the disease phenotype of the heterozygote more closely resembles that of the homozygote for two variant alleles. This phenomenon occurs in a number of trinucleotide repeat diseases: for an example and more details see Huntington Disease.

Dominant and Recessive Genetic Diseases in Humans

In humans, many genetic traits or diseases are classified simply as "dominant" or "recessive." Especially with respect to so-called recessive diseases, this can oversimplify the underlying molecular basis and lead to misunderstanding of the nature of dominance. For example, the genetic disease phenylketonuria (PKU) results from any of a large number (>60) of alleles at the gene locus for the enzyme phenylalanine hydroxylase (PAH). Many of these alleles produce little or no PAH, as a result of which the substrate phenylalanine and its metabolic byproducts accumulate in the central nervous system and can cause severe mental retardation if untreated.

The genotypes and phenotypic consequences of interactions among three alleles are shown in the following table:

Genotype	*PAH activity*	*[phe] conc*	*PKU ?*
AA	100%	60 uM	No
AB	30%	120 uM	No
CC	5%	200 ~ 300 uM	Hyperphenylalanemia
BB	0.3%	600 ~ 2400 uM	Yes

In unaffected persons homozygous for a standard functional allele (AA), PAH activity is standard (100%), and the concentration of phenylalanine in the blood [phe] is about 60 uM. In untreated persons homozygous for one of the PKU alleles (BB), PAH activity is close to zero, [phe] ten to forty times standard, and the individual manifests PKU.

In the AB heterozygote, PAH activity is only 30% (not 50%) of standard, blood [phe] is elevated two-fold, and the person does not manifest PKU. Thus, the A allele is dominant to the B allele with respect to PKU, but the B allele is incompletely dominant to the A allele with respect to its molecular effect, determination of PAH activity level (0.3% < 30% << 100%). Finally, the A allele is an incomplete dominant to B with respect to [phe], as 60 uM < 120 uM << 600 uM. Note once more that it is irrelevant to the question of dominance that the recessive allele produces a more extreme [phe] phenotype. For a third allele C, a CC homozygote produces a very small amount of PAH enzyme, which results in a somewhat elevated level of [phe] in the blood, a condition called hyperphenylalanemia, which does not result in mental retardation. That is, the dominance relationships of any two alleles may vary according to which aspect of the phenotype is under consideration. It is typically more useful to talk about the phenotypic consequences of the allelic interactions involved in any genotype, rather than to try to force them into dominant and recessive categories.

History

The concept of dominance was first described by the "Father of Genetics," Gregor Mendel, in the 1860s. Mendel observed that, for a variety of traits of garden peas having to do with the appearance of seeds, seed pods, and plant appearance, there occurred two discrete phenotypes: round vs wrinkled, or yellow vs green seeds, red vs white flowers, tall vs short plants, and so on. When bred separately, the plants always produced the same phenotypes, generation after generation. However, when lines with different phenotypes were crossed (interbred), one and only one of the parental phenotypes showed up in the offspring: green, or round, or red, or tall, and so on.

However, when these hybrid plants were crossed, the offspring plants showed the two original phenotypes, in a characteristic 3:1 ratio, with the more common type having the phenotype of the parental hybrid plants. Mendel reasoned that each of the parents in the first cross were homozygotes for different alleles (AA and aa), that each contributed one allele to the offspring, such that all of these hybrids were heterozygotes (Aa), and that one of the two alleles in the hybrid cross dominated expression of the other: A masked a. The final cross between two heterozygotes (Aa X Aa) would produce AA, Aa, and aa offspring in a 1:2:1 genotype ratio with the first three classes showing the "A" phenotype, and the last showing the "a" phenotype, thereby producing the 3:1 phenotype ratio.

Mendel did not use the terms gene, allele, phenotype, genotype, homozygote, and heterozygote, all of which were introduced afterward. He did introduce the notation of capital and lowercase letters for dominant and recessive alleles, respectively, still in use today.

Mitochondrial DNA

Mitochondrial DNA (mtDNA) is the DNA located in organelles called mitochondria, structures within eukaryotic cells that convert the chemical energy from food into a form that cells can use, adenosine triphosphate (ATP). Most other DNA present in eukaryotic organisms is found in the cell nucleus. Mitochondrial DNA can be regarded as the smallest chromosome, and was the first significant part of the human genome to be sequenced. In most species, including humans, mtDNA is inherited solely from the mother. The DNA sequence of mtDNA has been determined from a large number of organisms and individuals (including some organisms that are extinct), and the comparison of those DNA sequences represents a mainstay of phylogenetics, in that it allows biologists to elucidate the evolutionary relationships among species. It also permits an examination of the relatedness of populations, and so has become important in anthropology and field biology.

Replication

mtDNA is replicated by the DNA polymerase gamma complex which is composed of a 140 kDa catalytic DNA polymerase encoded by the POLG gene and a 55 kDa accessory subunit encoded by the POLG2 gene. During embryogenesis, replication of mtDNA is strictly down-regulated from the fertilized oocyte through the preimplantation embryo. At the blastocyst stage, the onset of mtDNA replication is specific to the cells of the trophectoderm. In contrast, the cells of the inner cell mass restrict mtDNA replication until they receive the signals to differentiate to specific cell types.

Origin

Nuclear and mitochondrial DNA are thought to be of separate evolutionary origin, with the mtDNA being derived from the circular genomes of the bacteria that were engulfed by the early ancestors of today's eukaryotic cells. This theory is called the endosymbiotic theory. Each mitochondrion is estimated to contain 2-10 mtDNA copies. In the cells of extant organisms, the vast majority of the proteins present in the mitochondria (numbering approximately 1500 different types in mammals) are coded for by nuclear DNA, but the genes for some of them, if not most, are thought to have originally been of bacterial origin, having since been transferred to the eukaryotic nucleus during evolution.

Mitochondrial Inheritance

In most multicellular organisms, mtDNA is inherited from the mother (maternally inherited). Mechanisms for this include simple dilution (an egg contains 100,000 to 1,000,000 mtDNA molecules, whereas a sperm contains only 100 to 1000), degradation of sperm mtDNA in the fertilized egg, and, at least in a few organisms, failure of sperm mtDNA to enter the egg. Whatever the mechanism, this single parent (uniparental) pattern of mtDNA inheritance is found in most animals, most plants and in fungi as well.

Female Inheritance

In sexual reproduction, mitochondria are normally inherited exclusively from the mother. The mitochondria in mammalian sperm are usually destroyed by the egg cell after fertilization. Also, most mitochondria are present at the base of the sperm's tail, which is used for propelling the sperm cells. Sometimes the tail is lost during fertilization. In 1999 it was reported that paternal sperm mitochondria (containing mtDNA) are marked with ubiquitin to select them for later destruction inside the embryo. Some in vitro fertilization techniques, particularly injecting a sperm into an oocyte, may interfere with this.

The fact that mitochondrial DNA is maternally inherited enables researchers to trace maternal lineage far back in time. (Y-chromosomal DNA, paternally inherited, is used in an analogous way to trace the agnate lineage.)

This is accomplished in humans by sequencing one or more of the hypervariable control regions (HVR1 or HVR2) of the mitochondrial DNA, as with a genealogical DNA test. HVR1 consists of about 440 base pairs.

These 440 base pairs are then compared to the control regions of other individuals (either specific people or subjects in a database) to determine maternal lineage. Most often, the comparison is made to the revised Cambridge Reference Sequence. Vilà et al. have published studies tracing the matrilineal descent of domestic dogs to wolves. The concept of the Mitochondrial Eve is based on the same type of analysis, attempting to discover the origin of humanity by tracking the lineage back in time.

Because mtDNA is not highly conserved and has a rapid mutation rate, it is useful for studying the evolutionary relationships - phylogeny - of organisms. Biologists can determine and then compare mtDNA sequences among different species and use the comparisons to build an evolutionary tree for the species examined.

Because mtDNA is transmitted from mother to child (both male and female), it can be a useful tool in genealogical research into a person's maternal line.

Male Inheritance

It has been reported that mitochondria can occasionally be inherited from the father in some species such as mussels. Paternally inherited mitochondria have additionally been reported in some insects such as fruit flies, honeybees, and periodical cicadas.

Evidence supports rare instances of male mitochondrial inheritance in some mammals as well. Specifically, documented occurrences exist for mice, where the male-inherited mitochondria was subsequently rejected. It has also been found in sheep, and in cloned cattle.

It has been found in a single case in a human male. While many of these cases involve cloned embryos or subsequent rejection of the paternal mitochondria, others document in vivo inheritance and persistence under lab conditions.

Structure

In humans (and probably in metazoans in general), 100-10,000 separate copies of mtDNA are usually present per cell (egg and sperm cells are exceptions). In mammals, each double-stranded circular mtDNA molecule consists of 15,000-17,000 base pairs. The two strands of mtDNA are differentiated by their nucleotide content with the guanine rich strand referred to as the heavy strand, and the cytosine rich strand referred to as the light strand.

The heavy strand encodes 28 genes, and the light strand encodes 9 genes for a total of 37 genes. Of the 37 genes, 13 are for proteins (polypeptides), 22 are for transfer RNA (tRNA) and two are for the small and large subunits of ribosomal RNA (rRNA). This pattern is also seen among most metazoans, although in some cases one or more of the 37 genes is absent and the mtDNA size range is greater. Even greater variation in mtDNA gene content and size exists among fungi and plants, although there appears to be a core subset of genes that are present in all eukaryotes (except for the few that have no mitochondria at all). Some plant species have enormous mtDNAs (as many as 2,500,000 base pairs per mtDNA molecule) but, surprisingly, even those huge mtDNAs contain the same number and kinds of genes as related plants with much smaller mtDNAs.

The genome of the mitochondrion of the cucumber (Cucumis sativus) is three entirely or largely autonomous circular chromosomes (lengths 1556, 84 and 45 kilobases).

Genes

Note: The gene content of mitochondrial DNA varies among species. Information in this section is specific to human mitochondrial DNA

Electron Transport Chain

The mitochondrial genome contains 13 protein-coding genes. Many of these genes encode proteins of the electron transport chain:

Category	*Genes*
NADH dehydrogenase (complex I)	MT-ND1, MT-ND2, MT-ND3, MT-ND4, MT-ND4L, MT-ND5, MT-ND6
Coenzyme Q - cytochrome c reductase/Cytochrome b (complex III)	MT-CYB
cytochrome c oxidase (complex IV)	MT-CO1, MT-CO2, MT-CO3
ATP synthase	MT-ATP6, MT-ATP8

rRNA

Mitochondrial rRNA is encoded by MT-RNR1 (12S) and MT-RNR2 (16S).

tRNA

The following genes encode tRNA.

Amino Acid	*3-Letter*	*1-Letter*	*MT DNA*
Alanine	Ala	A	MT-TA
Arginine	Arg	R	MT-TR
Asparagine	Asn	N	MT-TN
Aspartic acid	Asp	D	MT-TD
Cysteine	Cys	C	MT-TC
Glutamic acid	Glu	E	MT-TE
Glutamine	Gln	Q	MT-TQ
Glycine	Gly	G	MT-TG
Histidine	His	H	MT-TH
Isoleucine	Ile	I	MT-TI
Leucine	Leu	L	MT-TL1, MT-TL2
Lysine	Lys	K	MT-TK
Methionine	Met	M	MT-TM
Phenylalanine	Phe	F	MT-TF
Proline	Pro	P	MT-TP
Serine	Ser	S	MT-TS1, MT-TS2
Threonine	Thr	T	MT-TT
Tryptophan	Trp	W	MT-TW
Tyrosine	Tyr	Y	MT-TY
Valine	Val	V	MT-TV

Mutations

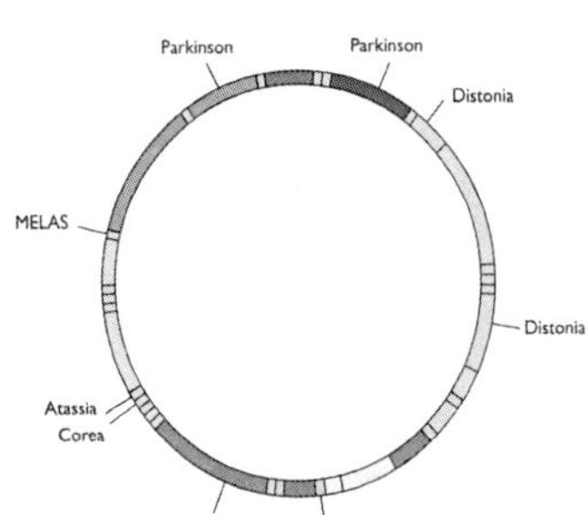

Figure: *The involvement of mitochondrial DNA in several human diseases.*

Susceptibility

mtDNA is particularly susceptible to reactive oxygen species generated by the respiratory chain due to its proximity. Though mtDNA is packaged by proteins and harbors significant DNA repair capacity, these protective functions are less robust than those operating on nuclear DNA and therefore thought to contribute to enhanced susceptibility of mtDNA to oxidative damage. The outcome of mutation in mtDNA may be alteration in the coding instructions for some proteins, which may have an effect on organism metabolism and/or fitness.

Genetic Illness

Mutations of mitochondrial DNA can lead to a number of illnesses including exercise intolerance and Kearns-Sayre syn drome (KSS), which causes a person to lose full function of heart, eye, and muscle movements. Some evidence suggests that they might be major contributors to the aging process and age-associated pathologies.

Use in Disease Diagnosis

Recently a mutation in mtDNA has been used to help diagnose prostate cancer in patients with negative prostate biopsy.

Use in Identification

In humans, mitochondrial DNA spans 16,569 DNA building blocks (base pairs), representing a fraction of the total DNA in cells. Unlike nuclear DNA, which is inherited from both parents and in which genes are rearranged in the process of recombination, there is usually no change in mtDNA from parent to offspring. Although mtDNA also recombines, it does so with copies of itself within the same mitochondrion. Because of this and because the mutation rate of animal mtDNA is higher than that of nuclear DNA, mtDNA is a powerful tool for tracking ancestry through females (matrilineage) and has been used in this role to track the ancestry of many species back hundreds of generations.

Human mtDNA can also be used to help identify individuals. Forensic laboratories occasionally use mtDNA comparison to identify human remains, and especially to identify older unidentified skeletal remains. Although unlike nuclear DNA, mtDNA is not specific to one individual, it can be used in combination with other evidence (anthropological evidence, circumstantial evidence, and the like) to establish identification. mtDNA is also used to exclude possible matches between missing persons and unidentified remains. Many researchers believe that mtDNA is better suited to identification of older skeletal remains than nuclear DNA because the greater number of copies of mtDNA per cell increases the chance of obtaining a useful sample, and because a match with a living relative is possible even if numerous maternal generations separate the two. American outlaw Jesse James's remains were identified using a comparison between mtDNA extracted from his remains and the mtDNA of the son of the female-line great-granddaughter of his sister. Similarly, the remains of Alexandra Feodorovna (Alix of Hesse), last Empress of Russia, and her children were identified by comparison of their mitochondrial DNA with that of Prince Philip, Duke of Edinburgh, whose maternal grandmother was Alexandra's sister Victoria of Hesse. Similarly to identify Emperor Nicholas II remains his mitochondrial DNA was compared with that of James Carnegie, 3rd Duke of Fife, whose maternal great-grandmother Alexandra of Denmark (Queen Alexandra) was sister of Nicholas II mother Dagmar of Denmark (Empress Maria Feodorovna).

The low effective population size and rapid mutation rate (in animals) makes mtDNA useful for assessing genetic relationships of individuals or groups within a species and also for identifying and quantifying the phylogeny (evolutionary relationships; phylogenetics) among different species, provided they are not too distantly related. To do this, biologists determine and then compare the mtDNA sequences from different individuals or species.

Data from the comparisons is used to construct a network of relationships among the sequences, which provides an estimate of the relationships among the individuals or species from which the mtDNAs were taken. This approach has limits that are imposed by the rate of mtDNA sequence change. In animals, the high mutation rate makes mtDNA most useful for comparisons of individuals within species and for comparisons of species that are closely or moderately-closely related, among which the number of sequence differences can be easily counted. As the species become more distantly related, the number of sequence differences becomes very large; changes begin to accumulate on changes until an accurate count becomes impossible.

History

Mitochondrial DNA was discovered in the 1960s by Margit M. K. Nass and Sylvan Nass by electron microscopy as DNase-sensitive thread inside mitochondria, and by Ellen Haslbrunner, Hans Tuppy and Gottfried Schatz by biochemical assays on highly purified mitochondrial fractions.

Sex Linkage

Sex linkage is the phenotypic expression of an allele related to the chromosomal sex of the individual. This mode of inheritance is in contrast to the inheritance of traits on autosomal chromosomes, where both sexes have the same probability of inheritance. Since humans have many more genes on the X than the Y, there are many more X-linked traits than Y-linked traits.

In mammals, the female is the homozygous sex, with two X chromosomes (XX), while the male is heterozygous, with one X and one Y chromosome (XY). Genes on the X or Y chromosome are called sex linked genes.

In birds, the opposite is true: the male is the homozygous sex, having two Z chromosomes (ZZ), and the female (hen) is heterozygous, having one Z and one W chromosome (ZW).

Linkage Overview

X-linked recessive traits are expressed in all heterogametics, but are only expressed in those homogametics that are homozygous for the recessive allele. For example, an X-linked recessive allele in humans causes hemophilia. Hemophilia is much more common in males than females because males are hemizygous - they only have one copy of the gene in question - and therefore express the trait when they inherit one mutant allele.

In contrast, a female must inherit two mutant alleles, a less frequent event since the mutant allele is rare in the population. Tsarevich Alexei of Russia was the most famous sufferer of X-linked hemophilia, and his disease may have played an important role in the overthrow of the imperial regime, which changed the course of history for millions of people.

The incidence of recessive X-linked phenotypes in females is the square of that in males (squaring a proportion less than one gives an outcome closer to 0 than the original). If 1 in 20 males in a human population are green colour blind, then 1 in 400 females in the population are expected to be colour blind $(/_{20})*(/_{20})$.

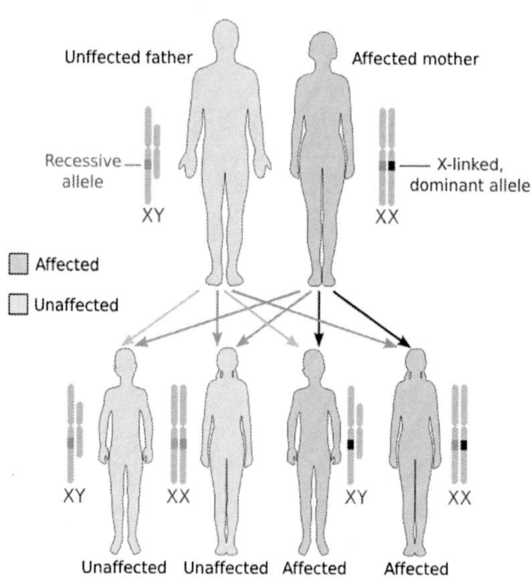

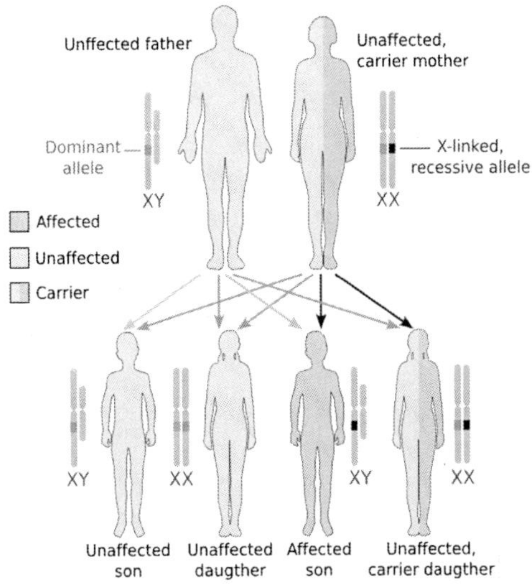

X-linked traits are maternally inherited from carrier mothers or from an affected father. Each son born to a carrier mother has a 50% probability of inheriting the X-chromosome carrying the mutant allele. There are a few Y-linked traits; these are inherited from the father.

In classical genetics, a reciprocal cross is performed to test if a trait is sex-linked.

Illustration of some X-linked heredity outcomes (A) the affected father has one X-linked dominant allele, the mother is homozygous for the recessive allele: only daughters (all) will be affected. (B) the affected mother is hemizygous with one copy of the X-linked dominant allele: both daughters and sons will have 50% probability to be affected. (C) the hemizygous mother is called "carrier" because she has one copy of the recessive allele: sons will have 50% probability to be affected, 50% of unaffected daughters will become carriers like their mother.

Autosomal Dominant Inheritance

Characteristics of Autosomal Dominant Inheritance: • The child of an affected parent has a 50% chance of inheriting the parent's mutated allele and thus being affected with the disorder. • A mutation can be transmitted by either the mother or the father. All children, regardless of gender, have an equal chance of inheriting the mutation

Examples:

• Tuberous sclerosis • Marfan syndrome • Neurofibromatosis • Huntington's disease • Retinoblastoma • Waardenburg syndrome • Myotonic dystrophy • Familial hypercholestrolemia (LDL receptor defect Type IIa) •

Adult polycystic kidney disease • von Hippel Lindau • Familial adenomatous polyposis and Peutz Jeghers Syndrome • Hereditory spherocytosis • Achondroplasia • Ehlor's Danlos (vascular type) • Acute intermittent porphyria • Hypertrophic Obstructive Cardiomyopathy (HOCM) • Von Willebrand Disease • Polydactyly • Osteogenesis Imperfecta (Except Type VII) • Hereditary hemorrhagic telengiactasia (Osler-weber-rendu syndrome) • Osteopetrosis Type II (Adult type) • Hypokalemic Periodic Paralysis

Autosomal Recessive Inheritance

Characteristics of Autosomal Recessive Inheritance: An individual will be a "carrier" if they posses one mutated allele and one normal gene copy. There is a 50% chance that a carrier will transmit a mutated gene to a child. If two carrier parents have a child there is a: o 25% chance that both will transmit the mutated gene; in this case, the child will inherit only mutated copies of the gene from both the mother and the father and thus will be affected with the disorder. o 50% chance that one carrier parent will transmit the mutated gene and the other will transmit the normal gene; in this case, the child will have one mutated gene and one normal gene and will be a carrier of the disorder. o 25% chance that both carrier parents will transmit the normal gene; in this case the child will have only normal genes and will not be affected and will not be a carrier.

Examples:

• cystic fibrosis, • sickle-cell disease, • Tay-Sachs disease, • Niemann-Pick disease, • spinal muscular atrophy, • Roberts's syndrome.

X-Linked Dominant Inheritance

Characteristics of X-Linked Dominant Inheritance: • A male or female child of an affected mother has a 50% chance of inheriting the mutation and thus being affected with the disorder. • All female children of an affected father will be affected (daughters possess their fathers' X-chromosome). • No male children of an affected father will be affected (sons do not inherit their fathers' X-chromosome).

Examples:

• Alport's syndrome • Aarsog's syndrome • Coffin - Lowry syndrome (CLS) • idiopathic hypoparathyroidism • incontinentia pigmenti • ornithine carbamoyl transferase deficiency • Rett syndrome (RS) • vitamin D resistant rickets • fragile X syndrome

X-Linked Recessive Inheritance

Characteristics of X-Linked Recessive Inheritance: • Females possessing one X-linked recessive mutation are considered carriers and will generally not manifest clinical symptoms of the disorder. • All males possessing an X-linked recessive mutation will be affected (males have a single X-chromosome and therefore have only one copy of X-linked genes). • All offspring of a carrier female have a 50% chance of inheriting the mutation. • All female children of an affected father will be carriers (daughters posses their fathers' X-chromosome). • No male children of an affected father will be affected (sons do not inherit their fathers' X-chromosome).

Examples:

• Lesch-Nyhan Syndrome • Duchene Muscular Dystrophy • Hunter's Disease • Menkes Disease (Kinky hair syndrome) • Glucose 6 Phosphate Dehydrogenase Deficiency • Hemophilia A and B • Fabry's Disease • Wiskott-Aldrich Syndrome • Bruton's Aggamaglobulinemia • Color Blindness • Complete Androgen Insensitivity • Congenital Aqueductal stenosis (hydrocephalus) • Inherited Nephrogenic Diabetes Insipidus

Y-linked

- Various failures in the SRY genes

Sex-linked Traits in Other aAnimals

- Calico or tortoiseshell cats, which have orange-and-black fur
- White eyes in Drosophila melanogaster flies—the first sex-linked gene ever discovered
- Evolution of dominance

The evolution of dominance concerns the evolution of genetic dominance. The central argument, that modifier genes act upon other genes to make them dominant or recessive, and that these are then themselves subject to natural selection was first proposed by the British population geneticist Ronald Fisher in 1928, and expanded upon in his book The Genetical Theory of Natural Selection. However, Sewall Wright and J.B.S. Haldane believed that the main explanation for dominance should be based on physiological factors, and that selection for modifiers was not a primary force. This led to a major fallout between Fisher and Wright. Subsequent works, particularly in molecular biology and biochemistry, have tended to favour Wright's view without completely excluding that the Fisherian argument may, in some circumstances, apply.

List of Mendelian Traits in Humans

In Mendelian inheritance, a child receiving a dominant allele from either parent will have the dominant form of the trait. Only those that received the recessive allele from both parents present with the recessive phenotype. Purely Mendelian traits are a tiny minority of all traits, since most phenotypic traits exhibit incomplete dominance, codominance, and contributions from many genes. The recessive phenotype may theoretically skip any number of generations, lying dormant in heterozygous "carrier" individuals until they have children with someone who also has the recessive allele and both pass it on to their child.

Examples

These traits include:

- Ability to taste phenylthiocarbamide (dominant)
- Ability to smell (bitter almond-like) hydrogen cyanide
- ACHOO syndrome (dominant)
- Albinism (recessive)
- Brachydactyly (Shortness of fingers and toes)
- Wet (dominant) or dry (recessive) earwax
- Face freckles (dominant)
- Huntingtons Disease (dominant)
- Blood type

Traits Previously Believed to be Mendelian

Some traits were previously believed to be Mendelian, but their inheritance is (probably) based on more complex genetic models , possibly involving more than one gene. These include :

- Eye color
- Hair color
- Morton's toe
- Tongue rolling
- Widow's peak (allele)
- Hitchhiker's thumb (recessive)
- Cleft chin (dominant)
- Dettached (dominant) or attached (recessive) earlobes

Autosomal Recessive

Two copies of the gene must be mutated for a person to be affected by an autosomal recessive disorder. An affected person usually has unaffected parents who each carry a single copy of the mutated gene (and are referred to as carriers). Two unaffected people who each carry one copy of the mutated gene have a 25% chance with each pregnancy of having a child affected by the disorder. Examples of this type of disorder are cystic fibrosis, sickle-cell disease, Tay-Sachs disease, Niemann-Pick disease, spinal muscular atrophy, and Roberts syndrome. Certain other phenotypes, such as wet versus dry earwax, are also determined in an autosomal recessive fashion.

X-linked Dominant

X-linked dominant inheritance, sometimes referred to as X-linked dominance, is a mode of genetic inheritance by which a dominant gene is carried on the X chromosome. As an inheritance pattern, it is less common than the X-linked recessive type. In medicine, X-linked dominant inheritance indicates that a gene responsible for a genetic disorder is located on the X chromosome, and only one copy of the allele is sufficient to cause the disorder when inherited from a parent who has the disorder.

X-linked dominant traits do not necessarily affect males more than females (unlike X-linked recessive traits). The exact pattern of inheritance varies, depending on whether the father or the mother has the trait of interest. All daughters of an affected father will also be affected but none of his sons will be affected (unless the mother is also affected). In addition, the mother of an affected son is also affected (but not necessarily the other way round).

X-linked dominant disorders are caused by mutations in genes on the X chromosome. Only a few disorders have this inheritance pattern, with a prime example being X-linked hypophosphatemic rickets. Males and females are both affected in these disorders, with males typically being more severely affected than females. Some X-linked dominant conditions such as Rett syndrome, incontinentia pigmenti type 2 and Aicardi syndrome are usually fatal in males either in utero or shortly after birth, and are therefore predominantly seen in females. Exceptions to this finding are extremely rare cases in which boys with Klinefelter syndrome (47,XXY) also inherit an X-linked dominant condition and exhibit symptoms more similar to those of a female in terms of disease severity.

The chance of passing on an X-linked dominant disorder differs between men and women. The sons of a man with an X-linked dominant disorder will all be unaffected (since they receive their father's Y chromosome), and his daughters will all inherit the condition. A woman with an X-linked dominant disorder has a 50% chance of having an affected fetus with each pregnancy, although it should be noted that in cases such as incontinentia pigmenti only female offspring are generally viable. In addition, although these conditions do not alter fertility per se, individuals with Rett syndrome or Aicardi syndrome rarely reproduce. Genetics As the X chromosome is one of the sex chromosomes (the other being the Y chromosome), X-linked inheritance is determined by the gender of the parent carrying a specific gene and can often seem complex. This is due to the fact that, typically, females have two copies of the X-chromosome, while males have only one copy. The difference between dominant and recessive inheritance patterns also plays a role in determining the chances of a child inheriting an X-linked disorder from their parentage. Males are normally hemizygous for the X chromosome, having only one copy. As a result, X-linked dominant disorders usually show higher expressivity in males than females.

Inheritance

X-linked dominant inheritance works differently depending upon whether the mother (left image) or father (right image) is the carrier of a gene that causes a disease or disorder. In X-linked dominant inheritance, when the mother alone is the carrier of a mutated, or defective gene associated with a disease or disorder; she herself will have the disorder. Her children will inherit the disorder as follows:

- Of her daughters and sons: 50% will have the disorder, 50% will be completely unaffected. Children of both genders have an even chance of receiving either of their mother's two X chromosomes, one of which contains the defective gene in question.

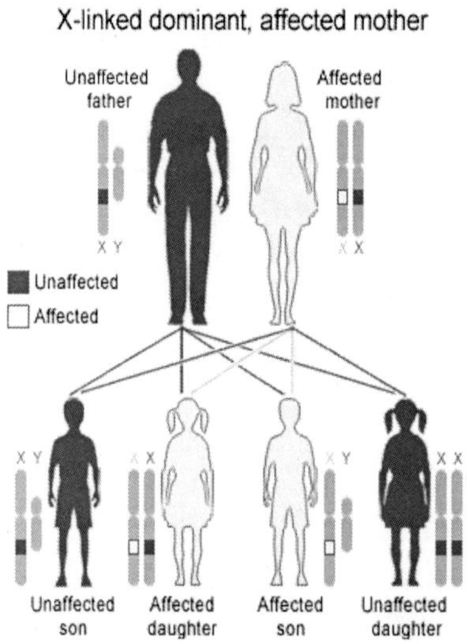

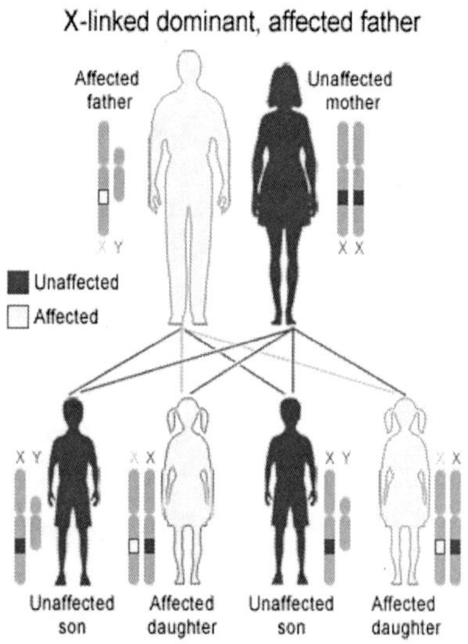

When the father alone is the carrier of a defective gene associated with a disease or disorder, he too will have the disorder. His children will inherit the disorder as follows:

- Of his daughters: 100% will have the disorder, since all of his daughters will receive one copy of his single X chromosome.
- Of his sons: none will have the disorder; sons do not receive an X chromosome from their father.

X-linked dominant, affected mother

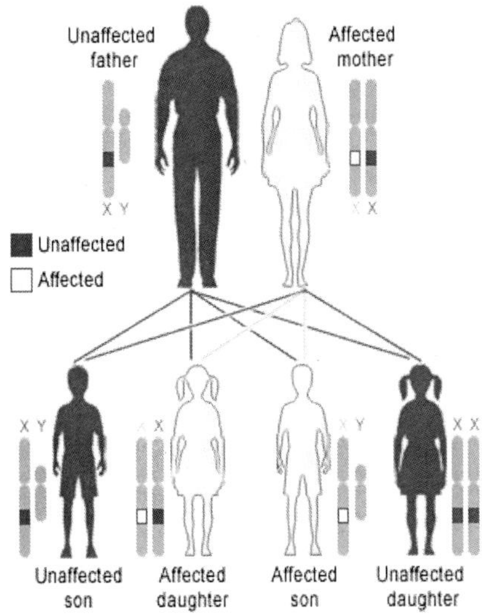

If both parents were carriers of a defective gene associated with a disease or disorder, they would both have the disorder. Their children would inherit the disorder as follows:

- Of their daughters: 100% will have the disorder, since all of the daughters will receive a copy of their father's X chromosome.
- Of the sons: 50% will have the disorder, 50% will be completely unaffected. Sons have an equal chance of receiving either of their mother's X chromosomes.

X-linked dominant, affected father

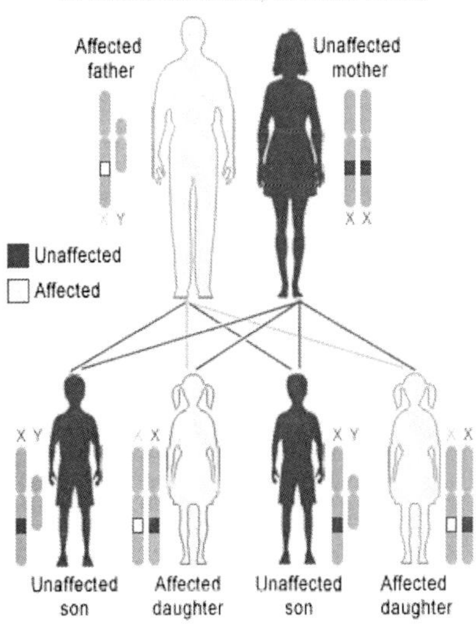

In such a case, where both parents carry and thus are affected by an X-linked dominant disorder, the chance of a daughter receiving two copies of the X chromosome with the defective gene is 50%, since daughters receive one copy of the X chromosome from both parents.

Were this to occur with an X-linked dominant disorder, that daughter would likely experience a more severe form. Some X-linked dominant conditions such as Aicardi Syndrome are fatal to boys, therefore only girls with these conditions survive. Similarly, individuals with Klinefelter's Syndrome are referred to as "47,XXY Males".

X-linked Hypophosphatemia

X-linked hypophosphatemia (XLH), also called X-linked dominant hypophosphatemic rickets, X-linked vitamin d-resistant rickets or hypophosphatemic vitamin d-resistant rickets (HPDR), is an X-linked dominant form of rickets (or osteomalacia) that differs from most cases of rickets in that ingestion of vitamin D is relatively ineffective.

It can cause bone deformity including short stature and genu varum (bow leggedness). It is associated with a mutation in the PHEX gene sequence (Xp.22) and subsequent inactivity of the PHEX protein. The prevalence of the disease is 1:20000. The leg deformity can be treated with Ilizarov frames and CHAOS surgery.

Cause and Genetics

XLH is associated with a mutation in the PHEX gene sequence, located on the human X chromosome at location Xp22.2-p22.1. The mutation results in altered (or missing) activity of the PHEX protein, which inactivates hormone-like substances (phosphatonins) that promote phosphate excretion. The resulting excess excretion of phosphate impairs bone mineralization. Biochemically, XLH is recognized by hypophosphatemia and inappropriately low level of calcitriol (1,25-$(OH)_2$ vitamin D_3).

The disorder is inherited in an X-linked dominant manner. This means the defective gene responsible for the disorder (PHEX) is located on the X chromosome, and only one copy of the defective gene is sufficient to cause the disorder when inherited from a parent who has the disorder. Males are normally hemizygous for the X chromosome, having only one copy. As a result, X-linked dominant disorders usually show higher expressivity in males than females.

As the X chromosome is one of the sex chromosomes (the other being the Y chromosome), X-linked inheritance is determined by the gender of the parent carrying a specific gene and can often seem complex. This is because, typically, females have two copies of the X-chromosome and males have only one copy. The difference between dominant and recessive inheritance patterns also plays a role in determining the chances of a child inheriting an X-linked disorder from their parentage.

X-linked Recessive

X-linked recessive conditions are also caused by mutations in genes on the X chromosome. Males are more frequently affected than females, and the chance of passing on the disorder differs between men and women. The sons of a man with an X-linked recessive disorder will not be affected, and his daughters will carry one copy of the mutated gene. A woman who is a carrier of an X-linked recessive disorder (X X) has a 50% chance of having sons who are affected and a 50% chance of having daughters who carry one copy of the mutated gene and are therefore carriers. X-linked recessive conditions include the serious diseases Hemophilia A, Duchenne muscular dystrophy, and Lesch-Nyhan syndrome as well as common and less serious conditions such as male pattern baldness and red-green color blindness. X-linked recessive conditions can sometimes manifest in females due to skewed X-inactivation or monosomy X (Turner syndrome).

Sex Differences In Phenotype/Genotypes and Frequency

In humans, generally "men are affected and women are carriers" for two reasons. The first is the simple statistical fact that if the X-chromosomes is a population that carry a particular X-linked mutation at a frequency of 'f' (for example, 1%) then that will be the frequency that men are likely to express the mutation (since they have only one X), while women will express it at a frequency of f (for example 1% * 1% = 0.01%) since they have two X's and hence two chances to get the normal allele. Thus, X-linked mutations tend to be rare in women.

Most Common

The most common X-linked recessive disorders are:

- Color blindness; a classic example of an X-linked trait because it is easy to explain the phenotype and it's relatively common 7% to 10% of men are red-green colorblind, making the above calculations predict 0.49% to 1% of women are carriers. Its commonness may be attributable to its not being a serious disability in most cases and an actual advantage in some situations (for example, not being distracted by some of the color in color based camouflage).It is also known as daltonism.
- Hemophilia A; another famous example because it was found in European royal families who intermarried and were famous enough that their

pedigrees could be established and offered in textbooks as a "famous example" of an X-linked trait that had been documented in history books before mendelian genetics was understood.

- Duchenne muscular dystrophy; muscular dystrophy associated with mutations in the dystrophin gene, characterized by rapid progression of muscle degeneration, eventually leading to loss in ambulation, respiratory failure and death.
- Becker's muscular dystrophy; milder form of Duchenne, causes slowly progressive muscle weakness of the legs and pelvis.
- Hemophilia B; a blood clotting disorder caused by a mutation of the Factor IX gene, leading to a deficiency of Factor IX. It is rarer than haemophilia A. It's also called Christmas disease
- X-linked ichthyosis; a skin condition (ichthyosis) caused by the hereditary deficiency of the steroid sulfatase (STS) enzyme that affects 1 in 2000 to 1 in 6000 males.
- X-linked agammaglobulinemia (XLA); affects the body's ability to fight infection (origin of the name: A=no, gammaglobulin=Antibody). XLA patients do not generate mature B cells. B cells are part of the immune system and normally manufacture antibodies (also called immunoglobulins) which defends the body from infections (the humoral response). Patients with untreated XLA are prone to develop serious and even fatal infections.
- Glucose-6-phosphate dehydrogenase deficiency; may exhibit nonimmune hemolytic anemia in response to a number of causes, most commonly infection or exposure to certain medications or chemicals.

Less Common Disorders

This section does not cite any references or sources. Please help improve this section by adding citations to reliable sources. Unsourced material may be challenged and removed. (March 2011)

Theoretically, a mutation in any of the genes in the Category:Genes on chromosome. X may cause disease, but below are some notable ones, with short description of symptoms:

- Adrenoleukodystrophy; leads to progressive brain damage, failure of the adrenal glands and eventually death.
- Alport syndrome; glomerulonephritis, endstage kidney disease, and hearing loss.
- Androgen insensitivity syndrome; variable degrees of undervirilization and/or infertility in XY persons of either gender.

- Barth syndrome; metabolism distortion, delayed motor skills, stamina deficiency, hypotonia, chronic fatigue, delayed growth, cardiomyopathy, and compromised immune system.
- Centronuclear myopathy; where cell nuclei are abnormally located in skeletal muscle cells. In CNM the nuclei are located at a position in the center of the cell, instead of their normal location at the periphery.
- Charcot-Marie-Tooth disease (CMTX2-3); disorder of nerves (neuropathy) that is characterized by loss of muscle tissue and touch sensation, predominantly in the feet and legs but also in the hands and arms in the advanced stages of disease.
- Coffin-Lowry syndrome; severe mental retardation sometimes associated with abnormalities of growth, cardiac abnormalities, kyphoscoliosis as well as auditory and visual abnormalities.
- Fabry disease; A lysosomal storage disease causing anhidrosis, fatigue, angiokeratomas, burning extremity pain and ocular involvement.
- Hunter's Syndrome; potentially causing hearing loss, thickening of the heart valves leading to a decline in cardiac function, obstructive airway disease, sleep apnea, and enlargement of the liver and spleen.
- Hypohidrotic ectodermal dysplasia, presenting with hypohidrosis, hypotrichosis, hypodontia
- Kabuki syndrome; multiple congenital anomalies and mental retardation.
- Kennedy disease; muscle cramps and progressive weakness
- Lesch-Nyhan syndrome; neurologic dysfunction, cognitive and behavioral disturbances including self-mutilation, and uric acid overproduction (hyperuricemia)
- Lowe Syndrome; hydrophthalmia, cataracts, intellectual disabilities, aminoaciduria, reduced renal ammonia production and vitamin D-resistant rickets
- Menkes disease; sparse and coarse hair, growth failure, and deterioration of the nervous system
- Nasodigitoacoustic syndrome; mishaped nose, brachydactyly of the distal phalanges, sensorineural deafness
- Nonsyndromic deafness and X-linked nonsyndromic deafness; hearing loss
- Occipital horn syndrome; deformations in the skeleton
- Ornithine transcarbamylase deficiency; developmental delay and mental retardation. Progressive liver damage, skin lesions, and brittle hair may also be seen

- Siderius X-linked mental retardation syndrome; cleft lip and palate with mental retardation and facial dysmorphism, caused by mutations in the histone demethylase PHF8
- Simpson-Golabi-Behmel syndrome; coarse faces with protruding jaw and tongue, widened nasal bridge, and upturned nasal tip
- Spinal muscular atrophy caused by UBE1 gene mutation; weakness due to loss of the motor neurons of the spinal cord and brainstem
- Wiskott-Aldrich syndrome; eczema, thrombocytopenia, immune deficiency, and bloody diarrhea
- X-linked Severe Combined Immunodeficiency (SCID); infections, usually causing death in the first years of life
- X-linked sideroblastic anemia; skin paleness, fatigue, dizziness and enlarged spleen and liver.

Y linkage

In mammals, Y-linkage refers to when a phenotypic trait is determined by an allele (or gene) on the Y chromosome. It is also known as holandric inheritance. The Y-chromosome is small and does not contain many genes, therefore few traits are Y-linked, and Y-linked diseases are rare. Because the only humans which have a Y chromosome are males, the genes are simply passed from father to son, with no interchromosomal genetic recombination.

Chromosome Y deletions are a frequent genetic cause of male infertility. Another example in humans of a y-linked trait was thought to be hairy ears (it may also be sex-limited). However, this has been discredited.

Y-linked

Y-linked disorders are caused by mutations on the Y chromosome. Because males inherit a Y chromosome from their fathers, every son of an affected father will be affected. Because females inherit an X chromosome from their fathers, female offspring of affected fathers are never affected. Since the Y chromosome is relatively small and contains very few genes, there are relatively few Y-linked disorders. Often the symptoms include infertility, which may be circumvented with the help of some fertility treatments. Examples are male infertility and hypertrichosis pinnae.

Mitochondrial

Mitochondrial diseases are a group of disorders caused by dysfunctional mitochondria, the organelles that are the "powerhouses" of the cell. Mitochondria are found in every cell of the human body except red blood cells. Mitochondria convert the energy of food molecules into the ATP that powers most cell functions.

Mitochondrial diseases are often caused by genetics or mutations to the mitochondrial DNA that affect mitochondria function. Mitochondrial diseases take on unique characteristics both because of the way the diseases are often inherited and because mitochondria are so critical to cell function.

The subclass of these diseases that have neuromuscular disease symptoms are often called a mitochondrial myopathy.

There is some evidence for mitochondrial dysfunction as the molecular basis of bipolar disorder. In addition, classical mitochondrial diseases occur in a subset of individuals with autism and are usually caused by genetic anomalies or mitochondrial respiratory pathway deficits.

This type of inheritance, also known as maternal inheritance, applies to genes in mitochondrial DNA. Because only egg cells contribute mitochondria to the developing embryo, only mothers can pass on mitochondrial conditions to their children. An example of this type of disorder is Leber's hereditary optic neuropathy.

Classification

In addition to the mitochondrial myopathies, other examples include:

- Diabetes mellitus and deafness (DAD)
  - this combination at an early age can be due to mitochondrial disease
  - Diabetes mellitus and deafness can also be found together for other reasons
- Leber's hereditary optic neuropathy (LHON)
  - visual loss beginning in young adulthood
  - eye disorder characterized by progressive loss of central vision due to degeneration of the optic nerves and retina
  - Wolff-Parkinson-White syndrome
  - multiple sclerosis-type disease
  - affects 1 in 50,000 people in Finland
- Leigh syndrome, subacute sclerosing encephalopathy
  - after normal development the disease usually begins late in the first year of life, although onset may occur in adulthood
  - a rapid decline in function occurs and is marked by seizures, altered states of consciousness, dementia, ventilatory failure
- Neuropathy, ataxia, retinitis pigmentosa, and ptosis (NARP)
  - progressive symptoms as described in the acronym
  - dementia

- Myoneurogenic gastrointestinal encephalopathy (MNGIE)
  - o gastrointestinal pseudo-obstruction
  - o neuropathy
- Myoclonic Epilepsy with Ragged Red Fibers (MERRF)
  - o progressive myoclonic epilepsy
  - o "Ragged Red Fibers" – clumps of diseased mitochondria accumulate in the subsarcolemmal region of the muscle fiber and appear as "Ragged Red Fibers" when muscle is stained with modified Gömöri trichrome stain
  - o short stature
  - o hearing loss
  - o lactic acidosis
  - o exercise intolerance
- Mitochondrial myopathy, encephalomyopathy, lactic acidosis, stroke-like symptoms (MELAS)
- mtDNA depletion
  - o mitochondrial neurogastrointestinal encephalomyopathy (MNGIE)

Nota bene: Conditions such as Friedreich's ataxia can affect the mitochondria, but are not associated with mitochondrial proteins.

Symptoms

Symptoms include poor growth, loss of muscle coordination, muscle weakness, visual problems, hearing problems, learning disabilities, heart disease, liver disease, kidney disease, gastrointestinal disorders, respiratory disorders, neurological problems, autonomic dysfunction, and dementia.

Characteristics

The effects of mitochondrial disease can be quite varied. Since the distribution of the defective mitochondrial DNA may vary from organ to organ within the body, and each mutation is modulated by other genome variants, the mutation that in one individual may cause liver disease might in another person cause a brain disorder. The severity of the specific defect may also be great or small. Some minor defects cause only "exercise intolerance", with no serious illness or disability. Defects often affect the operation of the mitochondria and multiple tissues more severely, leading to multi-system diseases. Mitochondrial diseases as a rule are worse when the defective mitochondria are present in the muscles, cerebrum, or nerves, because these cells use more energy than most other cells in the body.

Although mitochondrial diseases vary greatly in presentation from person to person, several major clinical categories of these conditions have been defined, based on the most common phenotypic features, symptoms, and signs associated with the particular mutations that tend to cause them.

An outstanding question and area of research is whether ATP depletion or reactive oxygen species are in fact responsible for the observed phenotypic consequences.

Causes

Mitochondrial disorders may be caused by mutations, acquired or inherited, in mitochondrial DNA (mtDNA) or in nuclear genes that code for mitochondrial components. They may also be the result of acquired mitochondrial dysfunction due to adverse effects of drugs, infections, or other environmental causes.

Mitochondrial DNA inheritance behaves differently from autosomal and sexually-linked inheritance. Nuclear DNA has two copies per cell (except for sperm and egg cells), one copy being inherited from the father and the other from the mother. Mitochondrial DNA, however, is strictly inherited from the mother and each mitochondrial organelle typically contains multiple mtDNA copies. During cell division the mitochondrial DNA copies segregate randomly between the two new mitochondria, and then those new mitochondria make more copies.

If only a few of the mtDNA copies inherited from the mother are defective, mitochondrial division may cause most of the defective copies to end up in just one of the new mitochondria (for more detailed inheritance patterns, see Human mitochondrial genetics). Mitochondrial disease may become clinically apparent once the number of affected mitochondria reaches a certain level; this phenomenon is called "threshold expression".

Mitochondrial DNA mutations occur frequently, due to the lack of the error checking capability that mitochondrial DNA has. This means that mitochondrial DNA disorders may occur spontaneously and relatively often. Defects in enzymes that control mitochondrial DNA replication (all of which are encoded for by genes in the nuclear DNA) may also cause mitochondrial DNA mutations.

Most mitochondrial function and biogenesis is controlled by nuclear DNA. Human mitochondrial DNA encodes only 13 proteins of the respiratory chain, while most of the estimated 1,500 proteins and components targeted to mitochondria are nuclear-encoded. Defects in nuclear-encoded mitochondrial genes are associated with hundreds of clinical disease phenotypes including anemia, dementia, hypertension, lymphoma, retinopathy, seizures, and neurodevelopmental disorders.

Treatment

Although research is ongoing, treatment options are currently limited; vitamins are frequently prescribed, though the evidence for their effectiveness is limited. Membrane penetrating antioxidants have the most important role in improving mitochondrial dysfunction.

Pyruvate has been proposed recently as a treatment option. Spindle transfer, where the nuclear DNA is transferred to another healthy egg cell leaving the defective mitochondrial DNA behind, is a potential treatment procedure that has been successfully carried out on monkeys. Using a similar pronuclear transfer technique, researchers at Newcastle University successfully transplanted healthy DNA in human eggs from women with mitochondrial disease into the eggs of women donors who were unaffected.

Human genetic engineering is already being used on a small scale to allow infertile women with genetic defects in their mitochondria to have children. Embryonic mitochondrial transplant and protofection have been proposed as a possible treatment for inherited mitochondrial disease, and allotopic expression of mitochondrial proteins as a radical treatment for mtDNA mutation load.

Statistics

About 1 in 4,000 children in the United States will develop mitochondrial disease by the age of 10 years. Up to 4,000 children per year in the US are born with a type of mitochondrial disease. Because mitochondrial disorders contain many variations and subsets, some particular mitochondrial disorders are very rare.

Many diseases of aging are caused by defects in mitochondrial function. Since the mitochondria are responsible for processing oxygen and converting substances from the foods we eat into energy for essential cellular functions, if there are problems with the mitochondria, it can lead to many defects for adults.

These include Type 2 diabetes, Parkinson's disease, atherosclerotic heart disease, stroke, Alzheimer's disease, and cancer. Many medicines can also injure the mitochondria. "Inside the Cell" in Dr. Neal Barnard's Program for Reversing Diabetes, Rodale Press, 2007, pp. 22 – 27, which references the Feb 12, 2004 issue of the New England Journal of Medicine, an article by Yale University researchers. Dr. Barnard also references other studies in his explanation of how, in Type 2 diabetes, the mitochondria signaling process is interrupted by fats in body cells (intramyocellular lipids) which have not been properly treated.

A study at Pennington Biomedical Research Cernter in Baton Brouge, LA (Diabetes 54, 2005 1926-33) showed that this in turn partially disables the genes that produce mitochondria.

People

Notable people who suffered from mitochondrial disease include:

- Rocco Baldelli (diagnosis later replaced by channelopathy)
- Mattie Stepanek (dysautonomic mitochondrial myopathy)
- Colby Wren

Multifactorial and Polygenic (Complex) Disorders

Genetic disorders may also be complex, multifactorial, or polygenic, meaning that they are likely associated with the effects of multiple genes in combination with lifestyle and environmental factors. Multifactorial disorders include heart disease and diabetes. Although complex disorders often cluster in families, they do not have a clear-cut pattern of inheritance. This makes it difficult to determine a person's risk of inheriting or passing on these disorders. Complex disorders are also difficult to study and treat because the specific factors that cause most of these disorders have not yet been identified. On a pedigree, polygenic diseases do tend to "run in families", but the inheritance does not fit simple patterns as with Mendelian diseases. But this does not mean that the genes cannot eventually be located and studied. There is also a strong environmental component to many of them (e.g., blood pressure).

- asthma
- autoimmune diseases such as multiple sclerosis
- cancers
- ciliopathies
- cleft palate
- diabetes
- heart disease
- hypertension
- inflammatory bowel disease
- mental retardation
- mood disorder
- obesity
- refractive error
- infertility.

3

Recent Developments in Stem Cell Technology

Neural stem cells can be cultured from the CNS of different mammalian species at many stages of development. They have an extensive capacity for self-renewal and will proliferate ex vivo in response to mitogenic growth factors or following genetic modification with immortalising oncogenes. Neural stem cells are multipotent since their differentiating progeny will give rise to the principal cellular phenotypes comprising the mature CNS: neurons, astrocytes and oligodendrocytes.

Neural stem cells can also be derived from more primitive embryonic stem (ES) cells cultured from the blastocyst. ES cells are considered to be pluripotent since they can give rise to the full cellular spectrum and will, therefore, contribute to all three of the embryonic germ layers: endoderm, mesoderm and ectoderm. However, pluripotent cells have also been derived from germ cells and teratocarcinomas (embryonal carcinomas) and their progeny may also give rise to the multiple cellular phenotypes contributing to the CNS. In a recent development, ES cells have also been isolated and grown from human blastocysts, thus raising the possibility of growing autologous stem cells when combined with nuclear transfer technology.

There is now an emerging recognition that the adult mammalian brain, including that of primates and humans, harbours stem cell populations suggesting the existence of a previously unrecognised neural plasticity to the mature CNS, and thereby raising the possibility of promoting endogenous neural reconstruction.

Such reports have fuelled expectations for the clinical exploitation of neural stem cells in cell replacement or recruitment strategies for the treatment of a variety of human neurological conditions including Parkinson's disease (PD), Huntington's disease, multiple sclerosis and ischaemic brain injury. Owing to their migratory capacity within the CNS, neural stem cells may also find potential clinical application as cellular vectors for widespread gene delivery and the expression of therapeutic proteins. In this regard, they may be eminently suitable for the correction of genetically-determined CNS disorders and in the management of certain tumours responsive to cytokines. Since large numbers of stem cells can be generated efficiently in culture, they may obviate some of the technical and ethical limitations associated with the use of fresh (primary) embryonic neural tissue in current transplantation strategies.

While considerable recent progress has been made in terms of developing new techniques allowing for the long-term culture of human stem cells, the successful clinical application of these cells is presently limited by our understanding of both (i) the intrinsic and extrinsic regulators of stem cell proliferation and (ii) those factors controlling cell lineage determination and differentiation. Although such cells may also provide accessible model systems for studying neural development, progress in the field has been further limited by the lack of suitable markers needed for the identification and selection of cells within proliferating heterogeneous populations of precursor cells. There is a further need to distinguish between the committed fate (defined during normal development) and the potential specification (implying flexibility of fate through manipulation of its environment) of stem cells undergoing differentiation.

With these challenges lying ahead, it is the opinion of the authors that stem-cell therapy is likely to remain within the experimental arena for the foreseeable future. In this regard, few (if any) of the in vivo studies employing neural stem cell grafts have shown convincingly that behavioural recovery can be achieved in the various model paradigms. Moreover, issues relating to the quality control of cultured cells and their safety following transplantation have only begun to be addressed.

While on the one hand cell biotechnologists have been quick to realise the potential commercial value, human stem cell research and its clinical applications has been the subject of intense ethical and legislative considerations.

During the past few weeks, significant developments have occurred that may strongly affect the progress of human stem cell research. Advances in both Somatic Cell Nuclear Transfer (SCNT) and in the so-called Yamanaka technique to produce patient-specific human pluripotent stem cells hold great promise not only for understanding disease development and progression, but for organ and tissue replacement. Unfortunately, progress has been delayed due to legal constraints that have restricted public access and widespread experimentation in many states of the nation. Although SCNT has worked effectively in experimental animals including mammals, it has not yet been used to produce patient-specific stem cells in humans. Last week, a research team led by Dr, Shoukhrat Mitalipov at the Oregon Health & Science University reported success using SCNT with rhesus macaque monkeys, the first report of success in using SCNT in primates.

It offers considerable promise and renewed hope that the procedural modification will work in other higher primates, including humans. The successful breakthrough was praised as "highly encouraging" by stem cell researchers far and wide, including Dr. Paul Simmons, Professor of Molecular Medicine at the University of Texas in Houston's Texas Medical Centre. Even the British creator of Dolly the cloned sheep, Prof. Ian Wilmut (Edinburgh University), praised the work of the Oregon team as a significant step forward toward human application. However, in a separate report, Prof. Wilmut personally vowed to abandon the SCNT technique for cloning human cells in favour of an emerging new procedure that utilises somatic cells (i.e., skin cells), rather than human blastocysts, the Yamanaka technique.

Dr. Wilmut was referring to a new stem cell technology created by a Japanese investigator that enables scientists to create pluripotent human stem cells from human somatic cells (i.e., skin fibroblasts) using a process known as reprogramming or dedifferentiation. The procedure, developed by Dr. Shinya Yamanaka, utilises a molecular biological approach in which cells are reprogrammed by exposing them to specific signaling agents (transcription factors) that reprogram the genetic expression pattern of a cell.

This prompts the cells to switch from becoming skin cells to becoming more stem cell-like. Although this approach is still in early stages, several seminal papers have just appeared in journals this week that give great credence to this alternative approach. Such developments should be seriously considered by the stem cell research community, as they offer yet another rational approach to making patient-specific, pluripotent stem cells. What is the Yamanaka technique and why is it considered such a breakthrough? To understand, let's first review some basic stem cell biology.

A stem cell is one that can replicate itself indefinitely while maintaining the ability to transform into other cell types. Stem cells presently come in two flavors: so-called adult and embryonic. Adult stem cells are found resident in most body organs while embryonic stem cells are present in a developing embryo for just a few days. Adult stem cells are said to be multipotent while embryonic stem cells are said to be pluripotent. The degree of potency conforms to the plasticity and degree of differentiation of the developing cell. At the time of the union of egg and sperm, the unified cell is said to be totipotent, meaning the cell can develop into any of the three types of tissue known as endoderm, mesoderm, or ectoderm as well as the placental tissues needed for the embryo to implant. As the cell progresses to two cells, then four, eight, and so forth, it reaches a stage where it is called a morula (berry in latin). About day four, the solid ball of cells begins to transform from a compressed morula into a hollowed out ball called a blastocyst. About the size of the period at the end of this sentence, this blastocyst contains a thin ridge of cells.

These are the embryonic stem cells. They were first isolated from and cultured in humans by James Thomson in 1998. These embryonic stem cells are said to be pluripotent. They can develop into any of the 210 or so cell lines of a human. As the blastocyst continues to develop the cells become more differentiated, more specialised. If the cell mass were divided before about day 14, it would create two viable identical cell masses. By day 14, the cells have differentiated to the point that if one were to try to divide them, the cell mass would arrest. By day 14, the cells have become so specialised, or differentiated, that they are no longer pluripotent, but are multipotent.

Adult stem cells have been used for over 40 years in such therapies as bone marrow transplantation. Embryonic stem cells have only been available for research and study for less than a decade. Adult stem cell research, whether the cells come from bone marrow, cord blood, peripheral blood, or aborted mesenchymal tissue, has not met with the controversy that has surrounded embryonic stem cell research. This is for one simple reason: the derivation of human embryonic stem cells.

Their derivation requires the destruction of human embryos. For this reason, George Bush issued a presidential proclamation in August, 2001 in which he proclaimed that the only federal funding for human embryonic stem cells would be for those embryonic stem cell lines already in existence. These became known as the presidential cell lines. His proclamation, combined with the Dickey amendment, effectively blocked any federal funding for human embryonic stem cell research.

The Dickey Amendment is the name of a piece of federal legislation passed by the United States Congress in 1995 which prohibits the Department of Health and Human Services (HHS) from using appropriated funds for the creation of human embryos for research purposes or for research in which human embryos are destroyed. HHS funding includes the funding for National Institutes of Health (NIH) funding. Technically the Dickey Amendment is a "rider" to other legislation, which amends the original legislation. The rider receives its name from the name of the Congressman that originally introduced the amendment, Representative Jay Dickey. The Dickey amendment language has been added to each of the Labor, HHS, and Education appropriations acts for FY1997 through FY2006. The original rider can be found in Section 128 of P.L. 104-99. The wording of the rider is generally the same year after year. For FY2007, the wording prohibits HHS from using FY2007 appropriated funds for:

1. the creation of a human embryo or embryos for research purposes; or
2. research in which a human embryo or embryos are destroyed, discarded, or knowingly subjected to risk of injury or death greater than that allowed for research on fetuses in utero under 45 CFR 46.208(a)(2) and Section 498(b) of the Public Health Service Act (42 U.S.C. 289g(b)) (Title 42, Section 289g(b), United States Code). For purposes of this section, the term "human embryo or embryos" includes any organism, not protected as a human subject under 45 CFR 46 (the Human Subject Protection regulations) ... that is derived by fertilization, parthenogenesis, cloning, or any other means from one or more human gametes (sperm or egg) or human diploid cells (cells that have two sets of chromosomes, such as somatic cells).

As a result of the federal embargo on funding human embryonic stem cell research, many of the 50 states have enacted various laws, propositions, and executive decrees that have created a patch quilt of legal and financial discord. This has impeded collaborative research. As well, key U.S. researchers have left the U.S. or have been recruited by foreign countries. The political fallout is part and parcel of the ethical debate, one fuelled by the means used to produce embryonic stem cells — the destruction of an embryo. To date, the only way to produce a human embryonic stem cell line is by extracting the little ridge of cells from the blastocyst.

SCNT, also called therapeutic cloning, has been used in mammals to create an embryonic stem cell line which is animal specific. Such a nuclear transfer was used to clone Dolly the sheep by Prof. Wilmut in 1995. It was the appeal of such a technique as SCNT that could generate patient-specific embryonic stem cells that allowed the Korean scientist Woo-Suk Hwang to captivate the world in 2006 with the promise of readily available, patient-specific human embryonic stem cells...until his work was shown to be fraudulent.

Last summer, at the annual ISSCR (International Society of Stem Cell Research) conference held in Australia, Dr. Shoukhrat Mitalipov and colleagues at the Oregon Health and Science University in Beaverton announced that they cloned the first nonhuman primate, a fact just recently appreciated by the general public with the pending publication of the scientific article. In all, Mitalipov used 304 eggs from 14 rhesus monkeys to make two lines of embryonic stem cells, one of which was chromosomally abnormal. Mitalipov admits the efficiency is low and, though his work is a proof of principle and the efficiency of his methods has improved, he admits it is not yet a cost-effective medical option. SCNT has been advanced by the advocates supporting human embryonic stem cell research as a means for producing a stem cell line without the destruction of a classically defined embryo.

In August, 2006 at the Stem Cell Conference in Keystone, Shinya Yamanaka, Professor of Stem Cell Biology at the Institute for Frontier Medical Sciences at Kyoto University in Japan, electrified the attending scientists by proving that it was possible to deprogram mouse somatic cells (specifically, fibroblasts) into embryonic-like stem cells by expressing four genes: c-myc, oct3/4, sox2, and klf4. According to Yamanaka, the proteins encoded by these genes, identified from a pool of 24 candidates, each serves a specific role. The Yamanaka technique offers a way to create human stem cells without destroying human embryos.

The significance of the work being published this week by two independent researchers is that for the first time the Yamanaka technique has been demonstrated as proof of concept in human beings. This process converts routine body cells, or somatic cells, into pluripotent stem cells by adding various transcription factors which cause reprogramming or dedifferentiation of the somatic cells. These reprogrammed somatic cells are called induced pluripotent stem cells or iPS cells.

These are just as plastic as embryonic stem cells. The one drawback to them is that they are created using virus vectors to introduce the genes. This is a major problem that must be solved before these cells can be used in humans. IPS cells have all the same characteristics of the pluripotent embryonic stem cells. They hold the same promise for use in regenerative medicine for cell replacement therapy.

Having been informed of these Richter-scale developments in creating pluripotent stem cells...the easy way, Prof Ian Wilmut, the father of modern therapeutic cloning is said to have "shunned" cloning. "I decided a few weeks ago (when he first heard about the studies) not to pursue nuclear transfer." He went on to say that the Yamanaka approach represents the future for stem cell research. Dr. Paul Simmons, immediate past president of the ISSCR, agrees with Prof Wilmut but suggests that "SCNT may still have a pulse, and is worth continued study." He goes on to point out that the long term comparisons of the iPS cells and embryonic stem cells still need to stand the test of time. "Embryonic stem cells derived from excess IVF embryos are still the gold standard for pluripotent stem cells. IPS cells are the new kids on the block. They need further study with comparisons made over time." Thomas Zwaka,

MD, PhD, assistant Professor of cell and molecular biology at the Baylor College of Medicine in Houston is more forceful in his opinion. As a long time collaborator of Yamanaka's and having spent four years as a post-doc fellow in James Thomson's lab, Zwaka — who attends to five presidential stem cell lines of his own — is well positioned to offer an overview. In his opinion, SCNT merits continued attention, while supporting active research utilising the Yamanaka technique.

"There is no excuse now for not funding this research," he said in an interview, "this is the beginning of a whole new frontier of medicine, a medicine of cures, not just treatments. Regenerative medicine will thrive using the cells derived from the Yamanaka technique." However, he echoes Paul Simmons in saying that other forms of stem cell derivation should be kept on reserve until all the kinks are worked out of the Yamanaka technique. He warns that "now is not the time to restrict any experimental approach, but to expand on all fronts until the right approach is confirmed."

Has the Yamanaka technique killed stem cell research as a political wedge issue in the upcoming elections? "No, probably not," says Bernard Siegel, executive director of the prestigious Genetics Policy Institute, a stem cell think tank based in Wellington, Florida whose advisory board reads like a who's who in stem cells. "Remember that the primary bone of contention has always been the use of embryos to derive the cell lines.

"The pro-life movement is still on the march, but with the Yamanaka technique, there's a different tune that appeals to all parties involved, Democrats and Republicans, right and left wings. The perennial argument over embryo use is muted, but still important. Now is the time to band together to get the cell replacement cures for spinal cord injury, Parkinson's, diabetes, and so many more disorders in place." Mr. Siegel, a successful attorney who has been a major leader in the advocacy movement for stem cell research, went on to point out that the Yamanaka procedure will offer both parties a compromise without the heated debate about the ethics and morality of using excess IVF embryos. "Stem cell research remains a potential wedge issue."

The following passage was written by Bill Brinkley, Ph.D., Professor of Cell and Molecular Biology, and Dean of the Graduate School of Houston's Baylor College of Medicine. "Although we are excited by these new discoveries, and remain even more optimistic about future developments of techniques that may utilise human stem cells for regenerative medicine, we believe that to disregard any procedure that currently holds promise is short-sighted and scientifically risky. A clear example is the Oregon team's development of the procedure that has worked effectively to produce a wide array of embryonic cells in non-human primates. To choose to focus on only one avenue of research or type of cell source, would — at this stage of regenerative medical research — be irresponsible, unreasonable, and premature. Promising and successful research exploring human stem cells should be supplemented with—not supplanted by — new and potentially exciting approaches, with all forms of research moving forward along multiple independent paths.

Scientific research — whether it be in cancer, diabetes, tissue regeneration or other areas — should proceed freely and openly along all viable lines of investigation until sufficient progress has been made to be successfully applied to the treatment and alleviation of diseases and human suffering.

In fact, these various lines of research will probably produce new findings that will complement each other and expand our depth and breadth of knowledge. Exciting new discoveries will be made in the field of stem cell research, and no one knows what important discoveries would be missed if we were to abandon SCNT to focus on only one approach, especially if that decision were largely driven by emotional and political expediency."

Medical Transplantation

Transplantation, medical, surgical procedure by which a tissue or organ is removed and replaced by a corresponding part, either from another part of the body or from another individual. A life-saving medical technique, transplantation also is an important tool in experimental biology; it is used to investigate endocrine gland functions, to study the interactions of cells in developing embryos, and to culture malignant tissue in cancer research.

Types of Transplanted Tissues and Organs

Transplantation to replace such diseased or defective tissue as corneas and hearts necessarily requires a dead donor; paired organs such as kidneys, or large or regenerating organs or tissues such as skin, bowel, lung, liver, or blood, can be donated by live donors. Skin autografts, employing skin from the patient's own body, are used to replace lost skin; autograft transplants are also done with bowel, bone, cartilage and other connective tissue, and ovarian tissue.

Replacement skin for autografts is now also grown in laboratories, and autograft bladders have been laboratory grown and implanted. Bone marrow transplants can come either from a donor or from stored host bone marrow. Controversial fetal tissue implants have been used for some neurodegenerative diseases and experimentally for fetus-to-fetus transplants in certain genetic disorders.

In addition to transplanted human tissues and organs, artificial parts ranging from heart valves to hip sockets are routinely implanted. Organ transplantation is the moving of an organ from one body to another or from a donor site on the patient's own body, for the purpose of replacing the recipient's damaged or absent organ. The emerging field of regenerative medicine is allowing scientists and engineers to create organs to be regrown from the patient's own cells (stem cells, or cells extracted from the failing organs). Organs and/or tissues that are transplanted within the same person's body are called autografts. Transplants that are performed between two subjects of the same species are called allografts. Allografts can either be from a living or cadaveric source.

Organs that can be transplanted are the heart, kidneys, eyes, liver, lungs, pancreas, intestine, and thymus. Tissues include bones, tendons (both referred to as musculoskeletal grafts), cornea, skin, heart valves, and veins. Worldwide, the kidneys are the most commonly transplanted organs, followed closely by the liver and then the heart. The cornea and musculoskeletal grafts are the most commonly transplanted tissues; these outnumber organ transplants by more than tenfold.

Organ donors may be living, or brain dead. Tissue may be recovered from donors who are cardiac dead – up to 24 hours past the cessation of heartbeat. Unlike organs, most tissues (with the exception of corneas) can be preserved and stored for up to five years, meaning they can be "banked". Transplantation raises a number of bioethical issues, including the definition of death, when and how consent should be given for an organ to be transplanted and payment for organs for transplantation. Other ethical issues include transplantation tourism and more broadly the socio-economic context in which organ harvesting or transplantation may occur. A particular problem is organ trafficking. Some organs, such as the brain, cannot be transplanted.

In the United States of America, tissue transplants are regulated by the U.S. Food and Drug Administration (FDA) which sets strict regulations on the safety of the transplants, primarily aimed at the prevention of the spread of communicable disease. Regulations include criteria for donor screening and testing as well as strict regulations on the processing and distribution of tissue grafts. Organ transplants are not regulated by the FDA.

Transplantation medicine is one of the most challenging and complex areas of modern medicine. Some of the key areas for medical management are the problems of transplant rejection, during which the body has an immune response to the transplanted organ, possibly leading to transplant failure and the need to immediately remove the organ from the recipient. When possible, transplant rejection can be reduced through serotyping to determine the most appropriate donor-recipient match and through the use of immunosuppressant drugs.

Autograft

Transplant of tissue to the same person. Sometimes this is done with surplus tissue, or tissue that can regenerate, or tissues more desperately needed elsewhere (examples include skin grafts, vein extraction for CABG, etc.) Sometimes an autograft is done to remove the tissue and then treat it or the person, before returning it (examples include stem cell autograft and storing blood in advance of surgery). In a rotationplasty a distal joint is used to replace a more proximal one, typically a foot and ankle joint is used to replace a knee joint. The a patient's foot is severed and reversed, the knee removed, and the tibia joined with the femur.

Allograft and Allotransplantation

An allograft is a transplant of an organ or tissue between two genetically non-identical members of the same species. Most human tissue and organ transplants are allografts. Due to the genetic difference between the organ and the recipient, the recipient's immune system will identify the organ as foreign and attempt to destroy it, causing transplant rejection.

Isograft

A subset of allografts in which organs or tissues are transplanted from a donor to a genetically identical recipient (such as an identical twin). Isografts are differentiated from other types of transplants because while they are anatomically identical to allografts, they do not trigger an immune response.

Xenograft and Xenotransplantation

A transplant of organs or tissue from one species to another. An example are porcine heart valve transplants, which are quite common and successful. Another example is attempted piscine-primate (fish to non-human primate) transplant of islet (i.e. pancreatic or insular tissue) tissue. The latter research study was intended to pave the way for potential human use, if successful. However, xenotransplantation is often an extremely dangerous type of transplant because of the increased risk of non-compatibility, rejection, and disease carried in the tissue.

Xenotransplantation

Xenotransplantation (*xenos-* from the Greek meaning "foreign") is the transplantation of living cells, tissues or organs from one species to another. Such cells, tissues or organs are called xenografts or xenotransplants. In contrast, the term allotransplantation refers to a same-species transplant. Human xenotransplantation offers a potential treatment for end-stage organ failure, a significant health problem in parts of the industrialised world. It also raises many novel medical, legal and ethical issues. A continuing concern is that many animals, such as pigs, have shorter lifespans than humans, meaning that their tissues age at a quicker rate. Disease transmission (xenozoonosis) and permanent alteration to the genetic code of animals are also causes for concern. There are few published cases of successful xenotransplantation.

Potential Future Uses

Because there is a worldwide shortage of organs for clinical implantation, about 60% of patients awaiting replacement organs die on the waiting list. Certain procedures, some of which are being investigated in early clinical trials, aim to use cells or tissues from other species to treat life-threatening and debilitating illnesses such as cancer, diabetes, liver failure and Parkinson's disease.

If vitrification can be perfected, it could allow for long-term storage of xenogenic cells, tissues and organs so that they would be more readily available for transplant. Xenotransplants could save thousands of patients waiting for donated organs. The animal organ, probably from a pig or baboon could be genetically altered with human genes to trick a patient's immune system into accepting it as a part of its own body. They have re-emerged because of the lack of organs available and the constant battle to keep immune systems from rejecting allotransplants. Xenotransplants are thus potentially a more effective alternative.

Also, xenotransplantation of ovarian tissue into immunodeficient nude mice or SCID mice is already used in research to study the development of ovarian follicles. Mature follicles have developed, even after use of cryopreserved ovarian tissue. Both host and graft vessels contribute to the revascularization of xenografted human ovarian tissue in a mice.

Potential Future Animal Organ Donors

Since they are the closest relatives to humans, nonhuman primates were first considered as a potential organ source for xenotransplantation to humans. Chimpanzees were originally considered to be the best option since their organs are of similar size, and they have good blood type compatibility with humans. However, since chimpanzees are listed as an endangered species, other potential donors were sought out. Baboons are more readily available, however they are also not practical as potential donors. Problems include their smaller body size, the infrequency of blood group O (the universal donor), their long gestation period, and they typically produce few offspring. In addition, a major problem with the use of nonhuman primates is the increased risk of disease transmission, since they are so closely related to humans. Pigs are currently thought to be the best candidates for organ donation. The risk of cross-species disease transmission is decreased because of their increased phylogenetic distance from humans . They are readily available, their organs are anatomically comparable in size, and new infectious agents are less likely since they have been in close contact with humans through domestication for many generations . Current experiments in xenotransplantation most often use pigs as the donor, and baboons as human models.

Barriers and Issues

Immunologic Barriers

To date no xenotransplantation trials have been entirely successful due to the many obstacles arising from the response of the recipient's immune system. This response, which is generally more extreme than in allotransplantations, ultimately results in rejection of the xenograft, and can in some cases result in the immediate death of the recipient. There are several types of rejection organ xenografts are faced with, these include:

- Hyperacute rejection
- Acute vascular rejection
- Cellular rejection
- Chronic rejection.

A rapid, violent hyperacute response results due to preformed natural antibodies, known as XNAs.

Hyperacute Rejection

This rapid and violent type of rejection occurs within minutes to hours from the time of the transplant. It is mediated by the binding of XNAs (xenoreactive natural antibodies) to the donor endothelium, causing activation of the human complement system which results in endothelial damage, inflammation, thrombosis and necrosis of the transplant. XNAs are first produced and begin circulating in the blood in neonates, after colonisation of the bowel by bacteria which have galactose moieties on their cell walls. Most of these antibodies are the IgM class, but also include IgG, and IgA. .

The epitope XNAs target is an α-linked galactose moiety, Gal-α-1,3Gal (also called the α-Gal epitope), produced by the enzyme α-galactosyl transferase. Most non-primates contain this enzyme thus, this epitope is present on the organ epithelium and is perceived as a foreign antigen by primates, which lack the galactosyl transferase enzyme. In pig to primate xenotransplantation, XNAs recognise porcine glycoproteins of the integrin family .

The binding of XNAs initiate complement activation through the classical complement pathway. Complement activation causes a cascade of events leading to: destruction of endothelial cells, platlet degranulation, inflammation, coagulation, fibrin deposition, and hemorrhage. The end result is thrombosis and necrosis of the xenograft .

Overcoming Hyperacute Rejection

Since hyperacute rejection presents such a barrier to the success of xenografts several strategies to overcome it are under investigation:

Interruption of the complement cascade:

- The recipient's complement cascade can be inhibited through the use of cobra venom factor (which depletes C3), soluble complement receptor type 1, anti-C5 antibodies, or C1 inhibitor (C1-INH). Disadvantages of this approach include the toxicity of cobra venom factor, and most importantly these treatments would deprive the individual of a functional complement system.

Transgeneic organs (Genetically engineered pigs):

- 1,3 galactosyl transferase gene knockouts - These pigs don't contain the gene which codes for the enzyme responsible for expression of the immunogeneic gal-α-1,3Gal moiety (the α-Gal epitope).
- Increased expression of H-transferase (α 1,2 fucosyltransferase), an enzyme that competes with galactosyl transferase. Experiments have shown this reduces α-Gal expression by 70%.
- Expression of human complement regulators (CD55, CD46, and CD59) to inhibit the complement cascade.
- Plasmaphoresis, on humans to remove 1,3 galactosyl-transferase, reduces the risk of activation of effector cells such as CTL (CD8 T cells), complement pathway activation and delayed type hypersensitivity (DTH).

Acute Vascular Rejection

Also known as delayed xenoactive rejection, this type of rejection occurs in discordant xenografts within 2 to 3 days, if hyperacute rejection is prevented. The process is much more complex than hyperacute rejection and is currently not completely understood. Acute vascular rejection requires de novo protein synthesis and is driven by interactions between the graft endothelial cells and host antibodies, macrophages, and platelets. The response is characterised by an inflammatory infiltrate of mostly macrophages and natural killer cells (with small numbers of T cells), intravascular thrombosis, and fibrinoid necrosis of vessel walls. Binding of the previously mentioned XNAs to the donor endothelium leads to the activation of host macrophages as well as the endothelium itself. The endothelium activation is considered type II since gene induction and protein synthesis are involved. The binding of XNAs ultimately leads to the development of a procoagulant state, the secretion of inflammatory cytokines and chemokines, as well as expression of leukocyte adhesion molecules such as E-selectin, intercellular adhesion molecule-1 (ICAM-1), and vascular cell adhesion molecule-1 (VCAM-1) . This response is further perpetuated as normally binding between regulatory proteins and their ligands aid in the control of coagulation and inflammatory responses. However, due to molecular incompatibilities between the molecules of the donor species and recipient (such as porcine major histocompatibility complex molecules and human natural killer cells), this may not occur.

Overcoming Acute Vascular Rejection

Due to its complexity, with the use of immunosuppressive drugs along with a wide array of approaches are necessary to prevent acute vascular rejection, and include:

- Administering a synthetic thrombin inhibitor to modulate thrombogenesis
- Depletion of anti-galactose antibodies (XNAs) by techniques such as immunoadsorption, to prevent endothelial cell activation
- Inhibiting activation of macrophages (stimulated by $CD4^+$ T cells) and NK cells (stimulated by the release of Il-2). Thus, the role of MHC molecules and T cell responses in activation would have to be reassessed for each species combo.

Accommodation

If hyperacute and acute vascular rejection are avoided accommodation is possible, which is the survival of the xenograft despite the presence of circulating XNAs. The graft is given a break from humoral rejection when the complement cascade is interrupted, circulating antibodies are removed, or their function is changed, or there is a change in the expression of surface antigens on the graft. This allows the xenograft to upregulate and express protective genes, which aid in resistance to injury, such as heme oxygenase-1 (an enzyme that catalyses the degradation of heme) .

Cellular Rejection

Rejection of the xenograft in hyperactute and acute vascular rejection is due to the response of the humoral immune system, since the response is elicited by the XNAs. Cellular rejection is based on cellular immunity, and is mediated by:

1. Natural killer cells, which accumulate in and damagee the xenograft; and

2. T-lymphocytes - which are activated by MHC molecules through both direct and indirect xenorecognition.

In direct xenorecognition, antigen presenting cells from the xenograft present peptides to recipient $CD4^+$ T cells via xenogeneic MHC class II molecules, resulting in the production of interleukin 2 (IL-2). Indirect xenorecognition involves the presentation of antigens from the xenograft by recipient antigen presenting cells to $CD4^+$ T cells. Antigens of phagocytosed graft cells can also be presented by the host's class I MHC molecules to $CD8^+$ T cells. The strength of cellular rejection in xenografts remains uncertain, however it is expected to be stronger than in allografts due to differences in peptides among different animals. This leads to more antigens potentially recognised as foreign, thus eliciting a greater indirect xenogenic response .

Overcoming Cellular Rejection

A proposed strategy to avoid cellular rejection is to induce donor non-responsiveness using haematopoietic chimerism. Donor stem cells are introduced into the bone marrow of the recipient, where they coexist with the recipient's stem cells. The bone marrow stem cells give rise to cells of all haematopoietic lineages, through the process of hematopoiesis. Lymphoid progenitor cells are created by this process and move to the thymus where negative selection eliminates T cells found to be reactive to self. The existence of donor stem cells in the recipient's bone marrow causes donor reactive T cells to be considered self and undergo apoptosis .

Chronic Rejection

This final type of rejection is slow and progressive, and is usually described in transplants which survive the initial rejection phases. Scientists are still unclear how chronic rejection exactly works, research in this area is difficult since xenografts rarely survive past the initial acute rejection phases. Nonetheless, it is known is that XNAs and the complement system are not primarily involved. Fibrosis in the xenograft occurs as a result of immune reactions, cytokines (which stimulate fibroblasts), or healing (following cellular necrosis in acute rejection). Perhaps the major cause of chronic rejection is arteriosclerosis. Lymphocytes, which were previously activated by antigens in the vessel wall of the graft, activate macrophages to secrete smooth muscle growth factors.

This results in a build up of smooth muscle cells on the vessel walls, causing the hardening and narrowing of vessels within the graft. Chronic rejection leads to pathologic changes of the organ, and is why transplants must be replaced after so many years. It is also anticipated that chronic rejection will be more aggressive in xenotransplants as opposed to allotransplants.

Physiology

Extensive research is required to determine whether animal organs can replace the physiological functions of human organs.

Many issues include:

- Size - Differences in organ size limit the range of potential recipients of xenotransplants.
- Longevity - The lifespan of most pigs is roughly 15 years, currently it is unknown whether or not a xenograft may be able to last longer than that.
- Hormone and protein differences - Some proteins will be molecularly incompatible, which could cause malfunction of important regulatory processes. These differences also make the prospect of hepatic xenotransplantation less promising, since the liver plays an important role in the production of so many proteins .
- Environment - For example, pig hearts work in a different anatomical site and under different hydrostatic pressure than in humans.
- Temperature - The body temperature of pigs is 39°C (2°C above the average human body temperature). Implications of this difference, if any, on the activity of important enzymes are currently unknown.

Xenozoonosis

Xenozoonosis, also known as zoonosis or xenosis, is the transmission of infectious agents between species via a xenograft. Animal to human infection is normally rare, but has occurred in the past. An example of such is the avian influenza, when an influenza A virus was passed from birds to humans. Xenotransplantation may increase the chance of disease transmission for 3 reasons: 1. Implantation breaches the physical barrier that normally helps to prevent disease transmission, 2. The recipient of the transplant will be severely immunosuppressed; and 3. Human complement regulators (CD46, CD55, and CD59) expressed in transgenic pigs have been shown to serve as virus receptors, and may also help to protect viruses from attack by the complement system.

Examples of viruses carried by pigs include porcine herpesvirus, rotavirus, parvovirus, and circovirus. Porcine herpesviruses and rotaviruses can be eliminated from the donor pool by screening, however others (such as parvovirus and circovirus) may contaminate food and footwear then re-infect the herd. Thus, pigs to be used as organ donors will have to be housed under strict regulations and screened regularly for microbes and pathogens. Unknown viruses, as well as those which aren't harmful in the animal, may also pose risks (Takeuchi and George, 2000). Of particular concern are PERVS (porcine endogenous retroviruses), vertically transmitted microbes which are imbedded in swine genomes. The risks with xenosis are twofold as not only could the individual become infected, but a novel infection could initiate an epidemic in the human population. Because of this risk, the FDA has suggested any recipients of xenotransplants shall be closely monitored for the remainder of their life, and quarantined if they show signs of xenosis.

Baboons and pigs carry myriad transmittable agents which are harmless in their natural host, but extremely toxic and deadly in humans. HIV is an example of a disease which is believed to have jumped from monkeys to humans. Researchers also do not know if an outbreak of infectious diseases could occur and if they could contain the outbreak even though they have measures for control. Another obstacle facing xenotransplants is that of the body's rejection of foreign objects by its immune system. These antigens (foreign objects) are often treated with powerful immunosuppressive drugs which could in turn make the patient vulnerable to other infections and actually aid the disease trying to be cured. This is the reason the organs would have to be altered to fit with the patients' DNA (histocompatibility).

In 2005, the Australian National Health and Medical Research Council (NHMRC) declared a eighteen-year moratorium on all animal-to-human transplantation, concluding that the risks of transmission of animal viruses to patients and the wider community had not been resolved. This was repealed in 2009 after an NHMRC review stated "... the risks, if appropriately regulated, are minimal and acceptable given the potential benefits.", citing international developments on the management and regulation of xenotransplantation by the World Health Organisation and the European Medicines Agency.

Porcine Endogenous Retroviruses

Endogenous retroviruses are remnants of ancient viral infections, found in the genomes of most, if not all, mammalian species. Integrated into the chromosomal DNA, they are vertically transferred through inheritance. Due to the many deletions and mutations they accumulate over time, they usually are not infectious in the host species, however the virus may become infectious in another species . PERVS were originally discovered as retrovirus particles released from cultured porcine kidney cells.

Most breeds of swine harbour approximately 50 PERV genomes in their DNA. Although it is likely that most of these are defective, some may be able to produce infectious viruses so every proviral genome must be sequenced to identify which ones pose a threat. In addition, through complementation and genetic recombination, two defective PERV genomes could give rise to an infectious virus: There are three subgroups of infectious PERVs (PERV-A,

PERV-B, and PERV-C). Experiments have shown that PERV-A and PERV-B can infect human cells in culture. To date no experimental xenotransplantations have demonstrated PERV transmission, yet this does not mean PERV infections in humans are impossible.

Ethicality

Xenografts have been a controversial procedure since they were first attempted. Many, including animal rights groups, strongly oppose killing animals in order to harvest their organs for human use. Religious beliefs, such as the Jewish and Muslim prohibition against eating pork, have been sometimes thought to be a problem, however according to a Council of Europe documentation both religions agree that this rule is overridden by the preservation of human life. Nevertheless, the prohibition of the consumption of pig, does pose problems in some Islamic communities. In general, the use of pig and cow tissue in humans has been met with little resistance, save some religious beliefs, a few philosophical objections and the occasional satirical barb. With the ability to now use these techniques now manifest, a precise moral and ethical doctrine from all major religions should be pronounced to keep the religious zealots out of the medical technological and testing regime. Experimentation without consent doctrines are now followed, which was not the case in the past, which may lead to new religious guidelines to further medical research on pronouced ecumenical guidelines. The "Common Rule" is the United States bio-ethics mandate as of 2011. Some ethical issues include informed consent complexities for research subjects, as well as the selection of human subjects, rights of patients and medical staff and public education (as many companies may go ahead with experiments without public awareness).

Split Transplants

Sometimes a deceased-donor organ, usually a liver, may be divided between two recipients, especially an adult and a child. This is not usually a preferred option because the transplantation of a whole organ is more successful.

Domino Transplants

This operation is performed on patients with cystic fibrosis because both lungs need to be replaced and it is a technically easier operation to replace the heart and lungs at the same time. As the recipient's native heart is usually healthy, it can be transplanted into someone else needing a heart transplant. This term is also used for a special form of liver transplant in which the recipient suffers from familial amyloidotic polyneuropathy, a disease where the liver slowly produces a protein that damages other organs. This patient's liver can be transplanted into an older patient who is likely to die from other causes before a problem arises.

This term also refers to a series of living donor transplants in which one donor donates to the highest recipient on the waiting list and the transplant centre utilises that donation to facilitate multiple transplants. These other transplants are otherwise impossible due to blood type or antibody barriers to transplantation. The "Good Samaritan" kidney is transplanted into one of the other recipients, whose donor in turn donates his or her kidney to an unrelated recipient. Depending on the patients on the waiting list, this has sometimes been repeated for up to six pairs, with the final donor donating to the patient at the top of the list. This method allows all organ recipients to get a transplant even if their living donor is not a match to them. This further benefits patients below any of these recipients on waiting lists, as they move closer to the top of the list for a deceased-donor organ. Johns Hopkins Medical Centre in Baltimore and Northwestern University's Northwestern Memorial Hospital have received significant attention for pioneering transplants of this kind.

Major Organs and Tissues Transplanted

Thoracic Organs

- Heart (Deceased-donor only)
- Lung (Deceased-donor and living-related lung transplantation)
- Heart/Lung (Deceased-donor and Domino transplant).

Abdominal Organs

- Kidney (Deceased-donor and Living-Donor)
- Liver (Deceased-donor and Living-Donor)
- Pancreas (Deceased-donor only)
- Intestine (Deceased-donor and Living-Donor)
- Stomach (Deceased-donor only)
- Testis.

Tissues, Cells, Fluids

- Hand (Deceased-donor only),
- Cornea (Deceased-donor only),
- Skin including Face replant (autograft) and Face transplant (extremely rare)
- Islets of Langerhans (Pancreas Islet Cells) (Deceased-donor and Living-Donor)
- Bone marrow/Adult stem cell (Living-Donor and Autograft)

- Blood transfusion/Blood Parts Transfusion (Living-Donor and Autograft)
- Blood vessels (Autograft and Deceased-Donor)
- Heart valve (Deceased-Donor, Living-Donor and Xenograft[Porcine/bovine])
- Bone (Deceased-Donor and Living-Donor).

Types of Donor

Organ donors may be living, or brain dead. Brain dead means the donor must have received an injury (either traumatic or pathological) to the part of the brain that controls heartbeat and breathing. Breathing is maintained via artificial sources, which, in turn, maintains heartbeat. Once brain death has been declared the person can be considered for organ donation. Criteria for brain death vary. Because less than 3% of all deaths in the U.S. are the result of brain death, the overwhelming majority of deaths are ineligible for organ donation, resulting in severe shortages. Tissue may be recovered from donors who are cardiac dead. That is, their breathing and heartbeat has ceased. They are referred to as cadaveric donors. In general, tissues may be recovered from donors up to 24 hours past the cessation of heartbeat. In contrast to organs, most tissues (with the exception of corneas) can be preserved and stored for up to five years, meaning they can be "banked." Also, more than 60 grafts may be obtained from a single tissue donor. Because of these three factors, the ability to recover from a non-heart beating donor, the ability to bank tissue, and the number of grafts available from each donor, tissue transplants are much more common than organ transplants. The American Association of Tissue Banks estimates that more than one million tissue transplants take place in the United States each year.

Living

In "living donors", the donor remains alive and donates a renewable tissue, cell, or fluid (e.g. blood, skin), or donates an organ or part of an organ in which the remaining organ can regenerate or take on the workload of the rest of the organ (primarily single kidney donation, partial donation of liver, small bowel). Regenerative medicine may one day allow for laboratory-grown organs, using patient's own cells via stem cells, or healthy cells extracted from the failing organs.

Deceased

Deceased (formerly cadaveric) are donors who have been declared brain-dead and whose organs are kept viable by ventilators or other mechanical mechanisms until they can be excised for transplantation. Apart from brain-stem dead donors, who have formed the majority of deceased donors for the last twenty years, there is increasing use of Donation after Cardiac Death Donors (formerly non-heart beating donors) to increase the potential pool of donors as demand for transplants continues to grow. These organs have inferior outcomes to organs from a brain-dead donor; however given the scarcity of suitable organs and the number of people who die waiting, any potentially suitable organ must be considered.

Reasons for Donation and Ethical Issues

Living Related Donors

Living related donors donate to family members or friends in whom they have an emotional investment. The risk of surgery is offset by the psychological benefit of not losing someone related to them, or not seeing them suffer the ill effects of waiting on a list.

Paired Exchange

A "paired-exchange" is a technique of matching willing living donors to compatible recipients using serotyping. For example a spouse may be willing to donate a kidney to their partner but cannot since there is not a biological match. The willing spouse's kidney is donated to a matching recipient who also has an incompatible but willing spouse.

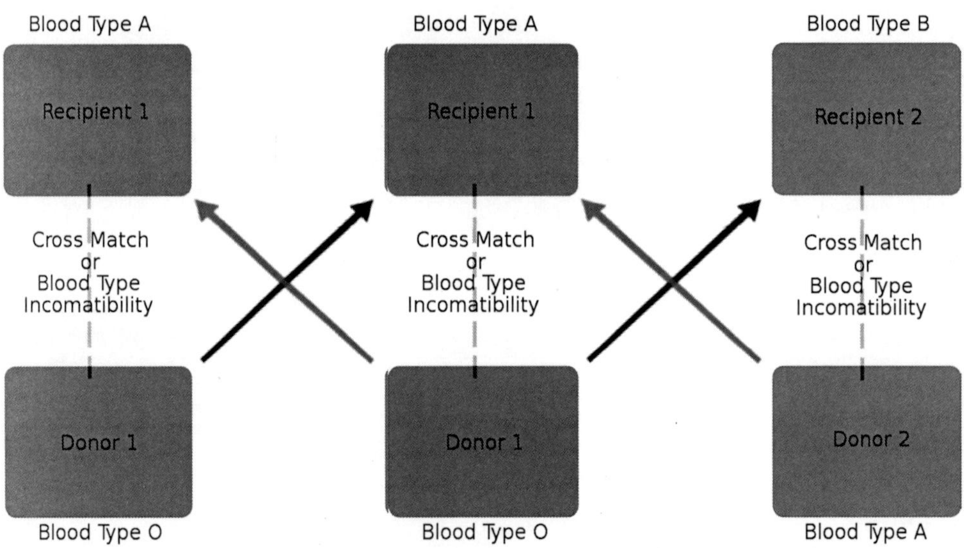

Figure: *Diagram of an exchange between otherwise incompatible pairs*

The second donor must match the first recipient to complete the pair exchange. Typically the surgeries are scheduled simultaneously in case one of the donors decides to back out and the couples are kept anonymous from each other until after the transplant.

Paired exchange programs were popularized in the New England Journal of Medicine article "Ethics of a paired-kidney-exchange program" in 1997 by L.F. Ross. It was also proposed by Felix T. Rapport in 1986 as part of his initial proposals for live-donor transplants "The case for a living emotionally related international kidney donor exchange registry" in *Transplant Proceedings.* A paired exchange is the simplest case of a much larger exchange registry program where willing donors are matched with any number of compatible recipients. Transplant exchange programs have been suggested as early as 1970: "A cooperative kidney typing and exchange program.".

The first pair exchange transplant in the U.S. was in 2001 at Johns Hopkins Hospital. The first complex multihospital kidney exchange involving 12 patients was performed in February 2009 by The Johns Hopkins Hospital, Barnes-Jewish Hospital in St. Louis and Integris Baptist Medical Centre in Oklahoma City. Another 12-patient multihospital kidney exchange was performed four weeks later by Saint Barnabas Medical Centre in Livingston, New Jersey, Newark Beth Israel Medical Centre and New York-Presbyterian Hospital. Surgical teams led by Johns Hopkins continue to pioneer in this field by having more complex chain of exchange such as eight-way multihospital kidney exchange. In December 2009, a 13 organ 13 recipient matched kidney exchange took place, coordinated through Georgetown University Hospital and Washington Hospital Centre, Washington DC.

Paired-donor exchange, led by work in the New England Program for Kidney Exchange as well as at Johns Hopkins University and the Ohio OPOs may more efficiently allocate organs and lead to more transplants.

Good Samaritan

Good Samaritan or "altruistic" donation is giving a donation to someone not well-known to the donor. Some people choose to do this out of a need to donate. Some donate to the next person on the list; others use some method of choosing a recipient based on criteria important to them. Web sites are being developed that facilitate such donation. It has been featured in recent television journalism that over half of the members of the Jesus Christians, an Australian religious group, have donated kidneys in such a fashion.

Compensated Donation

In compensated donation, donors get money or other compensation in exchange for their organs. This practice is common in some parts of the world, whether legal or not, and is one of the many factors driving medical tourism. In the United States, The National Organ Transplant Act of 1984 made organ sales illegal. In the United Kingdom, the Human Organ Transplants Act 1989 first made organ sales illegal, and has been superseded by the Human Tissue

Act 2004. In 2007, two major European conferences recommended against the sale of organs. Recent development of web sites and personal advertisements for organs among listed candidates has raised the stakes when it comes to the selling of organs, and have also sparked significant ethical debates over directed donation, "good-Samaritan" donation, and the current U.S. organ allocation policy. Bioethicist Jacob M. Appel has argued that organ solicitation on billboards and the internet may actually increase the overall supply of organs.

Two books, *Kidney for Sale By Owner* by Mark Cherry (Georgetown University Press, 2005); and *Stakes and Kidneys: Why markets in human body parts are morally imperative* by James Stacey Taylor: (Ashgate Press, 2005); advocate using markets to increase the supply of organs available for transplantation. In a 2004 journal article Economist Alex Tabarrok argues that allowing organ sales, and elimination of organ donor lists will increase supply, lower costs and diminish social anxiety towards organ markets.

Iran has had a legal market for kidneys since 1988, and the market price is of the order of US$1,200 for the recipient. *The Economist* and the Ayn Rand Institute approve and advocate a legal market elsewhere. They argued that if 0.06% of Americans between 19 and 65 were to sell one kidney, the national waiting list would disappear (which, the Economist wrote, happened in Iran). The Economist argued that donating kidneys is no more risky than surrogate motherhood, which can be done legally for pay in most countries.

In Pakistan, 40 percent to 50 percent of the residents of some villages have only one kidney because they have sold the other for a transplant into a wealthy person, probably from another country, said Dr. Farhat Moazam of Pakistan, at a World Health Organisation conference. Pakistani donors are offered $2,500 for a kidney but receive only about half of that because middlemen take so much. In Chennai, southern India, poor fishermen and their families sold kidneys after their livelihoods were destroyed by the Indian Ocean tsunami on December 26, 2004. About 100 people, mostly women, sold their kidneys for 40,000–60,000 rupees ($900–$1,350). Thilakavathy Agatheesh, 30, who sold a kidney in May 2005 for 40,000 rupees said, "I used to earn some money selling fish but now the post-surgery stomach cramps prevent me from going to work." Most kidney sellers say that selling their kidney was a mistake.

In Cyprus in 2010 police closed a fertility clinic under charges of trafficking in human eggs. The Petra Clinic, as it was known locally, imported women from Ukraine and Russia for egg harvesting and sold the genetic material to foreign fertility tourists. This sort of reproductive trafficking violates laws in the European Union. In 2010 the Pulitzer Centre on Crisis Reporting and the magazine Fast Company explored illicit fertility networks in Spain, the United States and Israel.

Allocation of Donated Organs

The overwhelming majority of deceased-donor organs in the United States are allocated by federal contract to the Organ Procurement and Transplantation Network (OPTN), held since it was created by the Organ Transplant Act of 1984 by the United Network for Organ Sharing or UNOS. (UNOS does not handle donor cornea tissue; corneal donor tissue is usually handled by various eye banks.) UNOS allocates organs based on the method considered most fair by the scientific leadership in the field. For kidneys, for instance, that is by waiting time; for livers, it is by MELD (Model of End-Stage Liver Disease), an empirical score based on lab values indicative of the sickness of the patient from liver disease. Experiencing somewhat increased popularity, but still very rare, is directed or targeted donation, in which the family of a deceased donor (often honouring the wishes of the deceased) requests an organ be given to a specific person.

If medically suitable, the allocation system is subverted, and the organ is given to that person. In the United States, there are various lengths of waiting due to the different availabilities of organs in different UNOS regions. In other countries such as the UK, only medical factors and the position on the waiting list can affect who receives the organ. If this is not the desired person, it is noted that this puts them higher on the list.

One of the more publicized cases of this type was the 1994 Chester and Patti Szuber transplant. This was the first time that a parent had received a heart donated by one of their own children. Although the decision to accept the heart from their recently killed child was not an easy decision, the Szuber family agreed that giving Patti's heart to her father would have been something that she would have wanted. Access to organ transplantation is one reason for the growth of medical tourism.

Forced Donation

There have been various accusations that certain authorities are harvesting organs from those the authorities deem undesirable, such as prison populations. The World Medical Association stated that individuals in detention are not in the position to give free consent to donate their organs . Illegal dissection of corpses is a form of body-snatching and may have taken place to obtain allografts.

According to the Chinese Deputy Minister of Health, Huang Jiefu, approximately 95% of all organs used for transplantation are from executed prisoners. The lack of public organ donation program in China is used as a justification for this practice. However reports in Chinese media raised concerns if executed criminals are the only source for organs used in transplants.

In October 2007, bowing to international pressure, the Chinese Medical Association agreed on a moratorium of commercial organ harvesting from condemned prisoners, but did not specify a deadline. China agreed to restrict transplantations from donors to their immediate relatives.

People in other parts of the world are responding to this availability of organs, and a number of individuals (including US and Japanese citizens) have elected to travel to China or India as medical tourists to receive organ transplants which may have been sourced in what might be considered elsewhere to be unethical ways.

Organ Transplantation in Different Countries

Demographics

Despite efforts of international transplantation societies, it is not possible to access an accurate source on the number, rates and outcomes of all forms of transplantation globally; the best that we can achieve is estimations. This is not a sound basis for the future and thus one of the crucial strategies for the Global Alliance in Transplantation is to foster the collection and analysis of global data.

All Numbers Per Million Population

According to the Council of Europe, Spain through the Spanish Transplant Organisation led by Dr Rafael Matesanz shows the highest worldwide rate of 35.1 donors per million population in 2005 and 33.8 in 2006. In addition to the citizens waiting for organ transplants in the US and other developed nations, there are long waiting lists in the rest of the world.

More than 2 million people need organ transplants in China, 50,000 waiting in Latin America (90% of which are waiting for kidneys), as well as thousands more in the less documented continent of Africa. Donor bases vary in developing nations. Traditionally, Muslims believe body desecration in life or death to be forbidden, and thus many reject organ transplant.

However most Muslim authorities nowadays accept the practice if another life will be saved. In Latin America the donor rate is 40–100 per million per year, similar to that of developed countries. However, in Uruguay, Cuba, and Chile, 90% of organ transplants came from cadaveric donors. Cadaveric donors represent 35% of donors in Saudi Arabia.

There is continuous effort to increase the utilisation of cadaveric donors in Asia, however the popularity of living, single kidney donors in India yields India a cadaveric donor prevalence of less than 1 pmp. Organ transplantation in China has taken place since the 1960s, and China has one of the largest transplant programmes in the world, peaking at over 13,000 transplants a year by 2004. Organ donation, however, is against Chinese tradition and culture, and involuntary organ donation is illegal under Chinese law. China's transplant programme attracted the attention of international news media in the 1990s due to ethical concerns about the organs and tissue removed from the corpses of executed criminals being commercially traded for transplants.

With regard to organ transplantation in Israel, there is a severe organ shortage due to religious objections by some rabbis who oppose all organ donations and others who advocate that a rabbi participate in all decision making regarding a particular donor. One third of all heart transplants performed on Israelis are done in the Peoples' Republic of China; others are done in Europe. Dr. Jacob Lavee, head of the heart-transplant unit, Sheba Medical Centre, Tel Aviv, believes that "transplant tourism" is unethical and Israeli insurers should not pay for it. The organisation HODS (Halachic Organ Donor Society) is working to increase knowledge and participation in organ donation among Jews throughout the world.

Transplantation rates also differ based on race, sex, and income. A study done with patients beginning long term dialysis showed that the sociodemographic barriers to renal transplantation present themselves even before patients are on the transplant list. For example, different groups express definite interest and complete pretransplant workup at different rates. Previous efforts to create fair transplantation policies had focused on patients currently on the transplantation waiting list.

Comparative Costs

One of the driving forces for illegal organ trafficking and for "transplantation tourism" is the price differences for organs and transplant surgeries in different areas of the world. According to the New England Journal of Medicine, a human kidney can be purchased in Manila for \$1000–\$2000, but in urban Latin America a kidney may cost more than \$10,000. Kidneys in South Africa have sold for as high as \$20,000. Price disparities based on donor race are a driving force of attractive organ sales in South Africa, as well as in other parts of the world.

In China, a kidney transplant operation runs for around \$70,000, liver for \$160,000, and heart for \$120,000. Although these prices are still unattainable to the poor, compared to the fees of the United States, where a kidney transplant may demand \$100,000, a liver \$250,000, and a heart \$860,000, Chinese prices have made China a major provider of organs and transplantation surgeries to other countries. In India, a kidney transplant operation runs for around as low as \$5000.

Safety

Compensation for donors also increases the risk of introducing diseased organs to recipients because these donors often yield from poorer populations unable to receive health care regularly and organ dealers may evade disease screening processes. The majority of such deals include one major payment and no follow up care for the donor. Some cases argue that there is a possibility of 1:18 to acquire HIV from such transplants.

In November 2007, the CDC reported the first-ever case of HIV and Hepatitis C being simultaneously transferred through an organ transplant. The donor was a 38-year-old male, considered "high-risk" by donation organisations, and his organs transmitted HIV and Hepatitis C to four organ recipients, none of whom had been told he was "high-risk." Experts say that the reason the diseases did not show up on screening tests is probably because they were contracted within three weeks before the donor's death, so antibodies would not have existed in high enough numbers to detect. The crisis has caused many to call for more sensitive screening tests, which could pick up antibodies sooner. Currently, the screens cannot pick up on the small number of antibodies produced in HIV infections within the last 90 days or Hepatitis C infections within the last 18–21 days before a donation is made.

NAT (nucleic acid testing) is now being done by many organ procurement organisations and is able to detect antibodies for HIV and Hepatitis C within seven to ten days of exposure to the virus.

Organ Transplant Laws

Both developing and developed countries have forged various policies to try to increase the safety and availability of organ transplants to their citizens. Brazil, France, Italy, Poland and Spain have ruled all adults potential donors with the "opting out" policy, unless they attain cards specifying not to be. However, whilst potential recipients in developing countries may mirror their more developed counterparts in desperation, potential donors in developing countries do not.

The Indian government has had difficulty tracking the flourishing organ black market in their country and have yet to officially condemn it. Other countries victimized by illegal organ trade have implemented legislative reactions. Moldova has made international adoption illegal in fear of organ traffickers. China has made selling of organs illegal as of July 2006 and claims that all prisoner organ donors have filed consent. However, doctors in other countries, such as the United Kingdom, have accused China of abusing its high capital punishment rate. Despite these efforts, illegal organ trafficking continues to thrive and can be attributed to corruption in healthcare systems, which has been traced as high up as the doctors themselves in China, Ukraine, and India, and the blind eye economically strained governments and health care programs must sometimes turn to organ trafficking. Some organs are also shipped to Uganda and the Netherlands. This was a main product in the triangular trade in 1934.

Starting on May 1, 2007, doctors involved in commercial trade of organs will face fines and suspensions in China. Only a few certified hospitals will be allowed to perform organ transplants in order to curb illegal transplants. Harvesting organs without donor's consent was also deemed a crime.

On June 27, 2008, Indonesian, Sulaiman Damanik, 26, pleaded guilty in Singapore court for sale of his kidney to CK Tang's executive chair, Mr Tang Wee Sung, 55, for 150 million rupiah (S$ 22,200). The Transplant Ethics Committee must approve living donor kidney transplants. Organ trading is banned in Singapore and in many other countries to prevent the exploitation of "poor and socially disadvantaged donors who are unable to make informed choices and suffer potential medical risks." Toni, 27, the other accused, donated a kidney to an Indonesian patient in March, alleging he was the patient's adopted son, and was paid 186 million rupiah (20,200 US). Upon sentence, both would suffer each, 12 months in jail or 10,000 Singapore dollars (7,600 US) fine.

In an article appearing in the Econ Journal Watch, April 2004. Economist Alex Tabarrok examined the impact of direct consent laws on transplant organ availability. Tabarrok found that social pressures resisting the use of transplant organs decreased over time as the opportunity of individual decisions increased. Tabarrok concluded his study suggesting that gradual elimination of organ donation restrictions and move to a free market in organ sales will increase supply of organs and encourage broader social acceptance of organ donation as a practice.

Ethical Concerns

The existence and distribution of organ transplantation procedures in developing countries, while almost always beneficial to those receiving them, raise many ethical concerns. Both the source and method of obtaining the organ to transplant are major ethical issues to consider, as well as the notion of distributive justice.

The World Health Organisation argues that transplantations promote health, but the notion of "transplantation tourism" has the potential to violate human rights or exploit the poor, to have unintended health consequences, and to provide unequal access to services, all of which ultimately may cause harm. Regardless of the "gift of life", in the context of developing countries, this might be coercive. There is also a powerful opposing view, that trade in organs, if properly and effectively regulated to ensure that the seller is fully informed of all the consequences of donation, is a mutually beneficial transaction between two consenting adults.

Even within developed countries there is concern that enthusiasm for increasing the supply of organs may trample on respect for the right to life. The question is made even more complicated by the fact that the "irreversibility" criterion for legal death cannot be adequately defined and can easily change with changing technology.

Artificial Organ Transplantation

Surgeons in Sweden performed the first implantation of a synthetic trachea in July 2011, for a 36-year-old patient who was suffering from cancer. Stem cells taken from the patient's hip were treated with growth factors and incubated on a plastic replica of his natural trachea.

Immunological Rejection of Transplanted Tissue

In transplanting complex organs (but not small tissue grafts), the larger blood vessels of the organ are surgically connected to those of the recipient. Connective tissue cells gradually link together the graft and host tissue. The main obstacle to successful transplantation is the rejection of foreign tissue by the host. Transplanted tissue from another individual (i.e., homograft, or allograft, tissue) contains antigens that stimulate an immune response from the host's lymphocytes. Homograft tissue is normally destroyed within a few weeks; the rejection mechanism is similar to that by which the body resists infection. The greater the number of foreign antigens on the donor organ, the more rapid and severe the rejection reactions.

Organs donated from one identical twin to another are usually viable because such organs are antigenically identical, but even organs transplanted between individuals who are fairly closely matched antigenically, such as siblings, have a good chance of being rejected. An antigenic typing system based on human lymphocyte antigens (HLA typing), pioneered by Jean Dausset in Paris and Rose Payne at Stanford Univ., has made it possible to identify histocompatibility and minimiminimis rejection. Today, most recipients of transplants are maintained on immunosuppressive drugs. The side-effects of such antirejection drugs, which can themselves be life threatening, include increased risk of infection, cancer, diabetes, and other conditions. In time, however, many patients develop a tolerance to the implanted organs, and some can eventually be weaned off the drugs. Researchers continue to study various ways to fool the immune system into accepting foreign tissues or to take advantage of the immune response. A new technique for nerve transplant begins with the patient taking immunosuppressive drugs, but after the patient's damaged nerves begin to grow and connect along the transplant, the drugs are discontinued and the immune system is allowed to destroy the transplanted nerve. Noncellular tissues or tissues where the donor cells are not important to the graft (e.g., bone and cartilage) can usually be successfully transplanted without rejection. In these transplants the grafts provide nonliving structural support within which the recipient's living cells gradually become established. Corneal transplants have a high success rate largely because there are so few blood vessels in the cornea that corneal antigens may never enter the host's system to stimulate an immune reaction. Bone-marrow transplants effectively bring their own immune system with them, often rejecting the new host, instead of the other way around, in a reaction known as graft-versus-host disease.

Implantation of artificial organs, such as artificial bone, is successful because such organs (prostheses) do not produce antigenic substances. Artificial joints made of stainless steel have been developed; newer implants have used nonrusting titanium joints with the midsection of bone substitute composed of lightweight polyethylene. Organ transplants from animals to humans are subject to hyperacute rejection, and transplantation of tissues from animals has been attempted for almost a century without much success. Some progress has been made, however, in circumventing the immune reaction. In one experimental approach, the tissues and organs of transgenic pigs, genetically engineered animals that have had human genes inserted, are combined with newly developed immunosuppressive drugs.

In a potential step toward a different approach to developing swine that could be used as a source of organs, researchers have cloned pigs in which a gene that causes rejection by the human immune system has been genetically disrupted. The endangered species status of chimpanzees, genetically closest animals to humans, has eliminated their use as donors. Although transplants from animals to humans, called xenotransplants, might benefit the thousands of patients waiting for human organs, the possibility that they could spread some unknown animal virus into the human population has caused concern and delayed research experimentation.

History

Human tissue grafting was first performed in 1870 by a Swiss surgeon, Jacques Reverdin. In 1912 the French surgeon Alexis Carrel developed methods of joining blood vessels that made the transplantation of organs feasible. He advanced this technique further and stimulated the use of transplantation in experimental biology. He also developed fluids and the means of circulating them so that transplanted tissues could be kept alive outside a living body in artificial media. Theoretical work by Jean Dausset, George Davis Snell and Baruj Benacerraf on the genetic basis of histocompatibility paved the way for practical applications. In the 1940s, Sir Peter Brian Medawar and Sir Macfarlane Burnet described foreign tissue rejection and acquired immunological tolerance, opening the way for transplant operations. The first successful identical twin transplant of a human kidney was made by Joseph E. Murray in 1954. The first human heart transplant was performed by the South African surgeon Christiaan Barnard in 1967; in 1968, Edward D. Thomas performed the first successful bone-marrow transplant between people who were not twins. In the following decades liver, kidney, heart, pancreas, bone-marrow, small intestines, and multiple organ transplants became more and more routine.

As transplantation has become more common and more successful, the demand for organs has risen dramatically. The development of heart transplantation has produced an ongoing re-examination of the traditional biological and legal definitions of death, because obtaining a healthy organ for transplantation depends in large part on the earliest possible establishment of the donor's death. More than 2,000 heart transplants per year were being performed in the United States by the late 1990s, with thousands of patients waiting for available hearts. In all, more than 64,000 people were waiting to receive new organs, including hearts, kidneys, livers, lungs, and pancreases. Many people carry organ donor cards, which indicate their wish to donate if they are killed in an accident, and many states require hospitals to request donation from the families of eligible donors. A side effect of the demand for donated organs has been the increasing use of lung and liver tissue, as well as kidneys, from live donors.

In the late 1990s surprising successes were achieved in transplanting body parts other than organs. Surgeons in France and the United States were able to transplant hands that became partly functional without rejection crises. In 2005 a French surgical team achieved a partial face transplant, replacing damage areas (nose, lips, and chin) of a woman's face with skin and underlying tissues from a dead donor. Although receiving less attention, successful transplants of knees, the trachea (windpipe), and the larynx have also been achieved. Such operations, called nonvital transplants, have become possible owing to improved surgical techniques, monitoring of rejection, and drug therapy. Still largely experimental, they must be approved by ethics committees before being undertaken, especially as the risk of taking immunosuppressive drugs may outweigh the benefits of the operation.

4

Embryonic Stem Cell

Embryonic stem cells (ES cells) are pluripotent stem cells derived from the inner cell mass of the blastocyst, an early-stage embryo. Human embryos reach the blastocyst stage 4–5 days post fertilization, at which time they consist of 50–150 cells. Isolating the embryoblast or inner cell mass (ICM) results in destruction of the fertilized human embryo, which raises ethical issues.

Characteristics

Embryonic stem cells are distinguished by two distinctive properties:

- their pluripotency, and
- their ability to replicate indefinitely.

ES cells are pluripotent, that is, they are able to differentiate into all derivatives of the three primary germ layers: ectoderm, endoderm, and mesoderm. These include each of the more than 220 cell types in the adult body. Pluripotency distinguishes embryonic stem cells from adult stem cells found in adults; while embryonic stem cells can generate all cell types in the body, adult stem cells are multipotent and can only produce a limited number of cell types. Human ES cells measure approximately 14ìm while mouse ES cells are closer to 8ìm.

Additionally, under defined conditions, embryonic stem cells are capable of propagating themselves indefinitely. This allows embryonic stem cells to be employed as useful tools for both research and regenerative medicine, because they can produce limitless numbers of themselves for continued research or clinical use. Because of their plasticity and potentially unlimited capacity for self-renewal, ES cell therapies have been proposed for regenerative medicine and tissue replacement after injury or disease. Diseases that could potentially be treated by pluripotent stem cells include a number of blood and immune-system related genetic diseases, cancers, and disorders; juvenile diabetes; Parkinson's; blindness and spinal cord injuries.

Besides the ethical concerns of stem cell therapy, there is a technical problem of graft-versus-host disease associated with allogeneic stem cell transplantation.

However, these problems associated with histocompatibility may be solved using autologous donor adult stem cells, therapeutic cloning, stem cell banks or more recently by reprogramming of somatic cells with defined factors (e.g. induced pluripotent stem cells). Other potential uses of embryonic stem cells include investigation of early human development, study of genetic disease and as in vitro systems for toxicology testing.

Research History and Developments

In 1964, researchers isolated a single type of cell from a teratocarcinoma, a tumor now known to be derived from a germ cell. These cells isolated from the teratocarcinoma replicated and grew in cell culture as a stem cell and are now known as embryonic carcinoma (EC) cells.

Although similarities in morphology and differentiating potential (pluripotency) led to the use of EC cells as the *in vitro* model for early mouse development, EC cells harbour genetic mutations and often abnormal karyotypes that accumulated during the development of the teratocarcinoma. These genetic aberrations further emphasised the need to be able to culture pluripotent cells directly from the inner cell mass.

In 1981, embryonic stem cells (ES cells) were independently first derived from mouse embryos by two groups. Martin Evans and Matthew Kaufman from the Department of Genetics, University of Cambridge published first in July, revealing a new technique for culturing the mouse embryos in the uterus to allow for an increase in cell number, allowing for the derivation of ES cells from these embryos. Gail R. Martin, from the Department of Anatomy, University of California, San Francisco, published her paper in December and coined the term "Embryonic Stem Cell".

She showed that embryos could be cultured *in vitro* and that ES cells could be derived from these embryos. In 1998, a breakthrough occurred when researchers, led by James Thomson at the University of Wisconsin-Madison, first developed a technique to isolate and grow human embryonic stem cells in cell culture.

Techniques and Conditions for Embryonic Stem Cell Derivation and Culture

Derivation of Human Embryonic Stem Cells

In vitro fertilization generates multiple embryos. The surplus of embryos is not clinically used or is unsuitable for implantation into the patient, and therefore may be donated by the donor with consent. Human embryonic stem cells are derived from these donated embryos that would otherwise be discarded.

The inner cell mass (cells of interest), from the blastocyst stage of the embryo, is separated from the trophectoderm, the cells that would differentiate into extra-embryonic tissue. Immunosurgery, the process in which antibodies are bound to the trophectoderm and removed by another solution, and mechanical dissection are performed to achieve separation.

The resulting inner cell mass cells are plated onto cells that will supply support. The inner cell mass cells attach and expand further to form a human embryonic cell line, which are undifferentiated. These cells are fed daily and are enzymatically or mechanically separated every four to seven days. For differentiation to occur, the human embryonic stem cell line is removed from the supporting cells to form embryoid bodies, is co-cultured with a serum containing necessary signals, or is grafted in a three-dimensional scaffold to result.

Derivation of Embryonic Stem Cells from other Animals

Embryonic stem cells are derived from the inner cell mass of the early embryo, which are harvested from the donor mother animal. Martin Evans and Matthew Kaufman reported a technique that delays embryo implantation, allowing the inner cell mass to increase.

This process includes removing the donor mother's ovaries and dosing her with progesterone, changing the hormone environment, which causes the embryos to remain free in the uterus. After 4–6 days of this intrauterine culture, the embryos are harvested and grown in *in vitro* culture until the inner cell mass forms "egg cylinder-like structures," which are dissociated into single cells, and plated on fibroblasts treated with mitomycin-c (to prevent fibroblast mitosis). Clonal cell lines are created by growing up a single cell. Evans and Kaufman showed that the cells grown out from these cultures could form teratomas and embryoid bodies, and differentiate *in vitro,* which all indicate the cells are pluripotent.

Gail Martin derived and cultured her ES cells differently. She removed the embryos from the donor mother at approximately 76 hours after copulation and cultured them overnight in media containing serum. The following day, she removed the inner cell mass from the late blastocyst using microsurgery.

The extracted inner cell mass was cultured on fibroblasts treated with mitomycin-c in media that containing serum and was conditioned by EC cells. After approximately one week, colonies of cells grew out. These cells grew in culture and demonstrated pluripotent characteristics, as demonstrated by the ability to form teratomas, differentiate *in vitro,* and form embryoid bodies. Martin referred to these cells as ES cells.

It is now known that the feeder cells provide leukemic inhibitory factor (LIF) and serum provides bone morphogenetic proteins (BMPs) that are necessary to prevent ES cells from differentiating. These factors are extremely important for the efficiency of deriving ES cells. Furthermore, it has been demonstrated that different mouse strains have different efficiencies for isolating ES cells. Current uses for mouse ES cells include the generation of transgenic mice, including knockout mice.

For human treatment, there is a need for patient specific pluripotent cells. Generation of human ES cells is more difficult and faces ethical issues. So, in addition to human ES cell research, many groups are focused on the generation of induced pluripotent stem cells (iPS cells).

Contamination by Reagents used in Cell Culture

The online edition of *Nature Medicine* published a study on January 24, 2005 which stated that the human embryonic stem cells available for federally funded research are contaminated with non-human molecules from the culture medium used to grow the cells. It is a common technique to use mouse cells and other animal cells to maintain the pluripotency of actively dividing stem cells. The problem was discovered when non-human sialic acid in the growth media was found to compromise the potential uses of the embryonic stem cells in humans, according to scientists at the University of California, San Diego.

However, a study published in the online edition of *Lancet Medical Journal* on March 8, 2005 detailed information about a new stem cell line which was derived from human embryos under completely cell- and serum-free conditions. After more than 6 months of undifferentiated proliferation, these cells demonstrated the potential to form derivatives of all three embryonic germ layers both *in vitro* and in teratomas. These properties were also successfully maintained (for more than 30 passages) with the established stem cell lines.

Potential Clinical Use

Reducing Donor-host Rejection

There is also ongoing research to reduce the potential for rejection of the differentiated cells derived from ES cells once researchers are capable of creating an approved therapy from ES cell research. One of the possibilities to prevent rejection is by creating embryonic stem cells that are genetically identical to the patient via therapeutic cloning. An alternative solution for rejection by the patient to therapies derived from non-cloned ES cells is to derive many well-characterised ES cell lines from different genetic backgrounds and use the cell line that is most similar to the patient; treatment can then be tailored to the patient, minimising the risk of rejection.

Safety: Reducing the Risk of Teratoma and Other Cancers as a Side Effect

- The major concern with the possible transplantation of ESC into patients as therapies is their ability to form tumours including teratoma. Safety issues prompted the FDA to place a hold on the first ESC clinical trial, however no tumours were observed.
- The main strategy to enhance the safety of ESC for potential clinical use is to differentiate the ESC into specific cell types (e.g. neurons, muscle, liver cells) that have reduced or eliminated ability to cause tumours. Following differentiation, the cells are subjected to sorting by flow cytometry for further purification. While ESC are predicted to be inherently safer than IPS cells because they are not genetically modified with genes such as c-Myc that are linked to cancer. Nonetheless ESC express very high levels of the iPS inducing genes and these genes including Myc are essential for ESC self-renewal and pluripotency, and potential strategies to improve safety by eliminating Myc expression are unlikely to preserve the cells' "stemness".

First Clinical Trial

On January 23, 2009, Phase I clinical trials for transplantation of oligodendrocytes (a cell type of the brain and spinal cord) derived from human ES cells into spinal cord-injured individuals received approval from the U.S. Food and Drug Administration (FDA), marking it the world's first human ES cell human trial.

The study leading to this scientific advancement was conducted by Hans Keirstead and colleagues at the University of California, Irvine and supported by Geron Corporation of Menlo Park, CA. A previous experiment had shown an improvement in locomotor recovery in spinal cord-injured rats after a 7 day delayed transplantation of human ES cells that had been pushed into an oligodendrocytic lineage.

In the proposed phase I clinical study, about eight to ten paraplegics who have had their injuries no longer than two weeks before the trial begins, will be selected, since the cells must be injected before scar tissue is able to form. However, the researchers are emphasising that the injections are not expected to fully cure the patients and restore all mobility. Based on the results of the rodent trials, researchers say restoration of myelin sheathes, and an increase in mobility is probable. This first trial is mainly testing the safety of these procedures and if everything goes well, it could lead to future studies that involve people with more severe disabilities. The trial had been put on hold in August 2009 due to concerns made by the FDA regarding a small number of microscopic cysts found in several treated rat models but the hold has been lifted as of July 30, 2010.

In October 2010 researchers enrolled and administered ESTs to the first patient at Shepherd Centre in Atlanta. The makers of the stem cell therapy, Geron Corporation, estimate that it will take several months for the stem cells to replicate and for the GRNOPC1 therapy to be evaluated for success or failure.

Potential Method for New cell Line Derivation

On August 23, 2006, the online edition of *Nature* scientific journal published a letter by Dr. Robert Lanza (medical director of Advanced Cell Technology in Worcester, MA) stating that his team had found a way to extract embryonic stem cells without destroying the actual embryo. This technical achievement would potentially enable scientists to work with new lines of embryonic stem cells derived using public funding in the USA, where federal funding was at the time limited to research using embryonic stem cell lines derived prior to August 2001. In March, 2009, the limitation was lifted.

Recently, it was shown that pluripotent stem cells highly similar to embryonic stem cells can be generated by the delivery of three genes (*Oct4*, *Sox2*, and *Klf4*) to differentiated cells. The delivery of these genes "reprograms" differentiated cells into pluripotent stem cells, allowing for the generation of pluripotent stem cells without the embryo. Because ethical concerns regarding embryonic stem cells typically are about their derivation from terminated embryos, it is believed that reprogramming to these "induced pluripotent stem cells" (iPS cells) may be less controversial.

Both human and mouse cells can be reprogrammed by this methodology, generating both human pluripotent stem cells and mouse pluripotent stem cells without an embryo.This may enable the generation of patient specific ES cell lines that could potentially be used for cell replacement therapies. In addition, this will allow the generation of ES cell lines from patients with a variety of genetic diseases and will provide invaluable models to study those diseases. However, as a first indication that the induced pluripotent stem cell (iPS) cell technology can in rapid succession lead to new cures, it was used by a research team headed by Rudolf Jaenisch of the Whitehead Institute for Biomedical Research in Cambridge, Massachusetts, to cure mice of sickle cell anemia, as reported by *Science* journal's online edition on December 6, 2007.

On January 16, 2008, a California based company, Stemagen, announced that they had created the first mature cloned human embryos from single skin cells taken from adults. These embryos can be harvested for patient matching embryonic stem cells.

Use of Human Embryonic Stem cells as Models for Human Genetic Disorders

Several new studies have started to address this issue. This has been done either by genetically manipulating the cells, or more recently by deriving diseased cell lines identified by prenatal genetic diagnosis (PGD). This approach may very well prove invaluable at studying disorders such as Fragile-X syndrome, Cystic fibrosis, and other genetic maladies that have no reliable model system.

Yury Verlinsky (Sept, 1, 1943 – July 16, 2009), a Russian-American medical researcher who specialised in embryo and cellular genetics (genetic cytology), developed prenatal diagnosis testing methods to determine genetic and chromosomal disorders a month and a half earlier than standard amniocentesis. The techniques are now used by many pregnant women and prospective parents, especially those couples with a history of genetic abnormalities or where the woman is over the age of 35, when the risk of genetically-related disorders is higher. In addition, by allowing parents to select an embryo without genetic disorders, they have the potential of saving the lives of siblings that already had similar disorders and diseases using cells from the disease free offspring.

Embryonic Stem Cell Research Oversight (ESCRO) Committees

The National Academies called for the establishment of Embryonic Stem Cell Research Oversight (ESCRO) Committees in its 2005 *Guidelines for Human Embryonic Stem Cell Research* to manage the ethical and legal concerns in human embryonic stem cell research. Because of the complexity and novelty of many of the issues involved in that research, the Guidelines committee believes that all research institutions engaged in human embryonic stem cell research should create and maintain these committees at the local level.

The composition and responsibilities of ESCRO committees was further clarified in the *Amendments to the National Academies' Guidelines for Human Embryonic Stem Cell Research*, released in February 2007.

Organisation

The National Academies' Guidelines make clear that activities related to human embryonic stem cell research should be overseen by an ESCRO committee. Those committees could be internal to a single institution or established jointly with one or more other institutions. Alternatively, an institution may have its proposals reviewed by an ESCRO committee of another institution or by an independent ESCRO committee. Many of these changes are discussed in the *2007 Amendments.*

Composition

The composition of ESCRO committees was specified to include representatives of the public and people with expertise in developmental biology, stem cell research, molecular biology, assisted reproduction, and ethical and legal issues in human embryonic stem cell research.

The *2007 Amendments* clarified that public representations should be independent and lay and in addition to the listed areas of scientific expertise. Although ESCRO committees may overlap with other oversight committees, it should not be a subcommittee of an Institutional Review Board, as its responsibilities extend beyond human subject protections.

Responsibilities

The Guidelines assigns several responsibilities to ESCRO committees:

- provide oversight over all issues related to derivation and use of hES cell lines
- review and approve the scientific merit of research protocols
- review compliance of all in-house human embryonic stem cell research with all relevant regulations and these guidelines
- maintain registries of human embryonic stem cell research conduced at the institution and hES cell lines derived or imported by institutional investigators
- facilitate education of investigators involved in human embryonic stem cell research.

Embryoid Body

Embryoid bodies are aggregates of cells derived from embryonic stem cells, and have been studied for years with mouse embryonic stem cells. Cell aggregation is imposed by hanging drop, plating upon non-tissue culture treated plates or spinner flasks; either method prevents cells from adhering to a surface to form the typical colony growth. Upon aggregation, differentiation is initiated and the cells begin to a limited extent to recapitulate embryonic development. Embryoid bodies are derived from pluripotent cells and therefore are made up of a large variety of differentiated cell types. However, the differentiation within the embryoid bodies, although occurring in a three dimensional manner is largely disorganised relative to the carefully orchestrated events of normal embryonic development. Still, embryoid bodies may serve as a good model system to investigate cellular and molecular interactions in the earliest stages of development which may be otherwise inaccessible for research, particularly in human.

The aggregate, while first simply appearing as a ball of cells, takes on an increasingly more complex appearance, becoming after a few days a hollow ball (cystic embryoid body), and next forms internal structures such as a yolk sac, and cardiomyocytes. Heart muscle cells which beat in a rhythmic pattern and neurons also commonly appear as part of the embyoid body.

Fetal Tissue Implant

Fetal tissue implant or fetal cell therapy is an experimental medical therapy where researchers implant tissue from a fetus to a patient for treatment of disease. In the case of Parkinson's disease, it is hoped that the fetal tissue would produce chemicals, specifically dopamine, which is lacking in the diseased brain. This therapy is also being investigated for treatment of Alzheimer's disease and Huntington's disease. Fetal tissue is unique since it is fast growing and has a lower possibility of rejection from the host's immune system than adult cells.

Uses

In 1982, seven people in Santa Clara County, California were diagnosed with Parkinsonism after having used MPPP contaminated with MPTP. In 1992, two of the seven patients were successfully treated at Lund University Hospital in Sweden with neural grafts of fetal tissue. One patient, who had been essentially paralysed, regained enough motor function to ride a bicycle.

Ethical and Political Concerns

Because the source of the tissue is aborted fetuses, there are significant legal and ethical issues being discussed. Federal funding for embryonic tissue research was restricted in the United States under Presidents Reagan and Bush before being lifted under the Clinton administration.

Adult Stem Cell

For many years, researchers have been seeking to understand the body's ability to repair and replace the cells and tissues of some organs, but not others. After years of work pursuing the how and why of seemingly indiscriminant cell repair mechanisms, scientists have now focused their attention on adult stem cells. It has long been known that stem cells are capable of renewing themselves and that they can generate multiple cell types. Today, there is new evidence that stem cells are present in far more tissues and organs than once thought and that these cells are capable of developing into more kinds of cells than previously imagined. Efforts are now underway to harness stem cells and to take advantage of this new found capability, with the goal of devising new and more effective treatments for a host of diseases and disabilities. What lies ahead for the use of adult stem cells is unknown, but it is certain that there are many research questions to be answered and that these answers hold great promise for the future.

What is an Adult Stem Cell?

Adult stem cells, like all stem cells, share at least two characteristics.

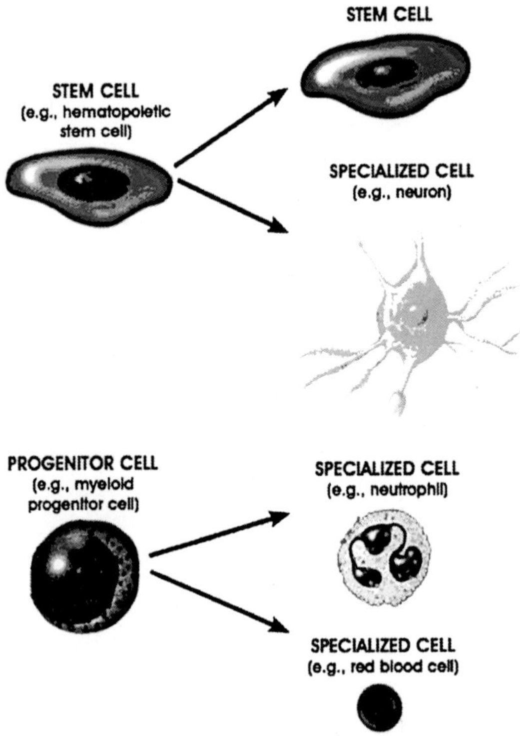

Figure: *A stem cell is an unspecialised cell that is capable of replicating or self renewing itself and developing into specialised cells of a variety of cell types. The product of a stem cell undergoing division is at least one additional stem cell that has the same capabilities of the originating cell. Shown here is an example of a hematopoietic stem cell producing a second generation stem cell and a neuron. A progenitor cell (also known as a precursor cell) is unspecialised or has partial characteristics of a specialised cell that is capable of undergoing cell division and yielding two specialised cells. Shown here is an example of a myeloid progenitor/ precursor undergoing cell division to yield two specialised cells (a neutrophil and a red blood cell).*

First, they can make identical copies of themselves for long periods of time; this ability to proliferate is referred to as long-term self-renewal. Second, they can give rise to mature cell types that have characteristic morphologies (shapes) and specialised functions. Typically, stem cells generate an intermediate cell type or types before they achieve their fully differentiated state. The intermediate cell is called a precursor or progenitor cell. Progenitor or precursor cells in fetal or adult tissues are partly differentiated cells that divide and give rise to differentiated cells. Such cells are usually regarded as "committed" to differentiating along a particular cellular development pathway, although this characteristic may not be as definitive as once thought.

Adult stem cells are rare. Their primary functions are to maintain the steady state functioning of a cell called homeostasis and, with limitations, to replace cells that die because of injury or disease. For example, only an estimated 1 in 10,000 to 15,000 cells in the bone marrow is a hematopoietic (bloodforming) stem cell (HSC). Furthermore, adult stem cells are dispersed in tissues throughout the mature animal and behave very differently, depending on their local environment. For example, HSCs are constantly being generated in the bone marrow where they differentiate into mature types of blood cells. Indeed, the primary role of HSCs is to replace blood cells. In contrast, stem cells in the small intestine are stationary, and are physically separated from the mature cell types they generate. Gut epithelial stem cells (or precursors) occur at the bases of crypts deep invaginations between the mature, differentiated epithelial cells that line the lumen of the intestine. These epithelial crypt cells divide fairly often, but remain part of the stationary group of cells they generate.

Unlike embryonic stem cells, which are defined by their origin (the inner cell mass of the blastocyst), adult stem cells share no such definitive means of characterisation. In fact, no one knows the origin of adult stem cells in any mature tissue. Some have proposed that stem cells are somehow set aside during fetal development and restrained from differentiating. Definitions of adult stem cells vary in the scientific literature range from a simple description of the cells to a rigorous set of experimental criteria that must be met before characterising a particular cell as an adult stem cell. Most of the information about adult stem cells comes from studies of mice. The list of adult tissues reported to contain stem cells is growing and includes bone marrow, peripheral blood, brain, spinal cord, dental pulp, blood vessels, skeletal muscle, epithelia of the skin and digestive system, cornea, retina, liver, and pancreas.

In order to be classified as an adult stem cell, the cell should be capable of self-renewal for the lifetime of the organism. This criterion, although fundamental to the nature of a stem cell, is difficult to prove *in vivo*. It is nearly impossible, in an organism as complex as a human, to design an experiment that will allow the fate of candidate adult stem cells to be identified *in vivo* and tracked over an individual's entire lifetime.

Ideally, adult stem cells should also be clonogenic. In other words, a single adult stem cell should be able to generate a line of genetically identical cells, which then gives rise to all the appropriate, differentiated cell types of the tissue in which it resides. Again, this property is difficult to demonstrate *in vivo*; in practice, scientists show either that a stem cell is clonogenic *in vitro*, or that a purified population of candidate stem cells can repopulate the tissue.

An adult stem cell should also be able to give rise to fully differentiated cells that have mature phenotypes, are fully integrated into the tissue, and are capable of specialised functions that are appropriate for the tissue. The term phenotype refers to all the observable characteristics of a cell (or organism); its shape (morphology); interactions with other cells and the non-cellular environment (also called the extracellular matrix); proteins that appear on the cell surface (surface markers); and the cell's behaviour (e.g., secretion, contraction, synaptic transmission).

The majority of researchers who lay claim to having identified adult stem cells rely on two of these characteristics appropriate cell morphology, and the demonstration that the resulting, differentiated cell types display surface markers that identify them as belonging to the tissue. Some studies demonstrate that the differentiated cells that are derived from adult stem cells are truly functional, and a few studies show that cells are integrated into the differentiated tissue in vivo and that they interact appropriately with neighbouring cells.

At present, there is, however, a paucity of research, with a few notable exceptions, in which researchers were able to conduct studies of genetically identical (clonal) stem cells. In order to fully characterize the regenerating and self-renewal capabilities of the adult stem cell, and therefore to truly harness its potential, it will be important to demonstrate that a single adult stem cell can, indeed, generate a line of genetically identical cells, which then gives rise to all the appropriate, differentiated cell types of the tissue in which it resides.

Evidence for the Presence of Adult Stem Cells

Adult stem cells have been identified in many animal and human tissues. In general, three methods are used to determine whether candidate adult stem cells give rise to specialised cells. Adult stem cells can be labelled *in vivo* and then they can be tracked. Candidate adult stem cells can also be isolated and labelled and then transplanted back into the organism to determine what becomes of them. Finally, candidate adult stem cells can be isolated, grown *in vitro* and manipulated, by adding growth factors or introducing genes that help determine what differentiated cells types they will yield. For example, currently, scientists believe that stem cells in the fetal and adult brain divide and give rise to more stem cells or to several types of precursor cells, which give rise to nerve cells (neurons), of which there are many types.

It is often difficult if not impossible to distinguish adult, tissue-specific stem cells from progenitor cells, which are found in fetal or adult tissues and are partly differentiated cells that divide and give rise to differentiated cells. These are cells found in many organs that are generally thought to be present to replace cells and maintain the integrity of the tissue. Progenitor cells give rise to certain types of cells such as the blood cells known as T lymphocytes, B lymphocytes, and natural killer cells but are not thought to be capable of developing into all the cell types of a tissue and as such are not truly stem cells. The current wave of excitement over the existence of stem cells in many adult tissues is perhaps fuelling claims that progenitor or precursor cells in those tissues are instead stem cells. Thus, there are reports of endothelial progenitor cells, skeletal muscle stem cells, epithelial precursors in the skin and digestive system, as well as some reports of progenitors or stem cells in the pancreas and liver. A detailed summary of some of the evidence for the existence of stem cells in various tissues and organs is presented later in the chapter.

Adult Stem Cell Plasticity

It was not until recently that anyone seriously considered the possibility that stem cells in adult tissues could generate the specialised cell types of another type of tissue from which they normally reside either a tissue derived from the same embryonic germ layer or from a different germ layer. For example, studies have shown that blood stem cells (derived from mesoderm) may be able to generate both skeletal muscle (also derived from mesoderm) and neurons (derived from ectoderm). That realisation has been triggered by a flurry of papers reporting that stem cells derived from one adult tissue can change their appearance and assume characteristics that resemble those of differentiated cells from other tissues.

The term plasticity, as used in this report, means that a stem cell from one adult tissue can generate the differentiated cell types of another tissue. At this time, there is no formally accepted name for this phenomenon in the scientific literature. It is variously referred to as "plasticity", "unorthodox differentiation" or "transdifferentiation".

Approaches for Demonstrating Adult Stem Cell Plasticity

To be able to claim that adult stem cells demonstrate plasticity, it is first important to show that a cell population exists in the starting tissue that has the identifying features of stem cells. Then, it is necessary to show that the adult stem cells give rise to cell types that normally occur in a different tissue. Neither of these criteria is easily met. Simply proving the existence of an adult stem cell population in a differentiated tissue is a laborious process. It requires that the candidate stem cells are shown to be self-renewing, and that they can give rise to the differentiated cell types that are characteristic of that tissue.

To show that the adult stem cells can generate other cell types requires them to be tracked in their new environment, whether it is *in vitro* or *in vivo*. In general, this has been accomplished by obtaining the stem cells from a mouse that has been genetically engineered to express a molecular tag in all its cells. It is then necessary to show that the labelled adult stem cells have adopted key structural and biochemical characteristics of the new tissue they are claimed to have generated. Ultimately and most importantly, it is necessary to demonstrate that the cells can integrate into their new tissue environment, survive in the tissue, and function like the mature cells of the tissue.

In the experiments reported to date, adult stem cells may assume the characteristics of cells that have developed from the same primary germ layer or a different germ layer. For example, many plasticity experiments involve stem cells derived from bone marrow, which is a mesodermal derivative. The bone marrow stem cells may then differentiate into another mesodermally derived tissue such as skeletal muscle, cardiac muscle or liver.

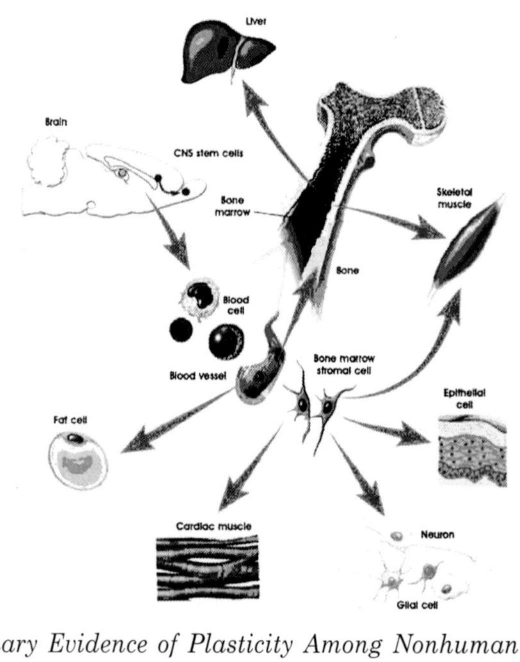

Figure: *Preliminary Evidence of Plasticity Among Nonhuman Adult Stem Cells.*

Alternatively, adult stem cells may differentiate into a tissue that during normal embryonic development would arise from a different germ layer. For example, bone marrow-derived cells may differentiate into neural tissue, which is derived from embryonic ectoderm. And reciprocally neural stem cell lines cultured from adult brain tissue may differentiate to form hematopoietic cells, or even give rise to many different cell types in a chimeric embryo. In both cases cited above, the cells would be deemed to show plasticity, but in the case of bone marrow stem cells generating brain cells, the finding is less predictable.

In order to study plasticity within and across germ layer lines, the researcher must be sure that he/she is using only one kind of adult stem cell. The vast majority of experiments on plasticity have been conducted with adult stem cells derived either from the bone marrow or the brain. The bone marrow-derived cells are sometimes sorted using a panel of surface markers into populations of hematopoietic stem cells or bone marrow stromal cells. The HSCs may be highly purified or partially purified, depending on the conditions used. Another way to separate population of bone marrow cells is by fractionation to yield cells that adhere to a growth substrate (stromal cells) or do not adhere (hematopoietic cells).

To study plasticity of stem cells derived from the brain, the researcher must overcome several problems. Stem cells from the central nervous system (CNS), unlike bone marrow cells, do not occur in a single, accessible location. Instead, they are scattered in three places, at least in rodent brain the tissue around the lateral ventricles in the forebrain, a migratory pathway for the cells that leads from the ventricles to the olfactory bulbs, and the hippocampus.

Many of the experiments with CNS stem cells involve the formation of neurospheres, round aggregates of cells that are sometimes clonally derived. But it is not possible to observe cells in the centre of a neurosphere, so to study plasticity *in vitro*, the cells are usually dissociated and plated in monolayers. To study plasticity *in vivo*, the cells may be dissociated before injection into the circulatory system of the recipient animal, or injected as neurospheres.

What is the Evidence for Plasticity?

The differentiated cell types that result from plasticity are usually reported to have the morphological characteristics of the differentiated cells and to display their characteristic surface markers. In reports that transplanted adult stem cells show plasticity *in vivo*, the stem cells typically are shown to have integrated into a mature host tissue and assumed at least some of its characteristics. Many plasticity experiments involve injury to a particular tissue, which is intended to model a particular human disease or injury. However, there is limited evidence to date that such adult stem cells can generate mature, fully functional cells or that the cells have restored lost function *in vivo*. Most of the studies that show the plasticity of adult stem cells involve cells that are derived from the bone marrow or brain. To date, adult stem cells are best characterised in these two tissues, which may account for the greater number of plasticity studies based on bone marrow and brain. Collectively, studies on plasticity suggest that stem cell populations in adult mammals are not fixed entities, and that after exposure to a new environment, they may be able to populate other tissues and possibly differentiate into other cell types.

It is not yet possible to say whether plasticity occurs normally in vivo. Some scientists think it may, but as yet there is no evidence to prove it. Also, it is not yet clear to what extent plasticity can occur in experimental settings, and how or whether the phenomenon can be harnessed to generate tissues that may be useful for therapeutic transplantation.

If the phenomenon of plasticity is to be used as a basis for generating tissue for transplantation, the techniques for doing it will need to be reproducible and reliable. In some cases, debate continues about observations that adult stem cells yield cells of tissue types different than those from which they were obtained.

Experimental Evidence of Adult Stem Cells and Plasticity

Adult Stem Cells of the Nervous System

More than 30 years ago, Altman and Das showed that two regions of the postnatal rat brain, the hippocampus and the olfactory bulb, contain dividing cells that become neurons. Despite these reports, the prevailing view at the time was that nerve cells in the adult brain do not divide. In fact, the notion that stem cells in the adult brain can generate its three major cell types astrocytes and oligodendrocytes, as well as neurons was not accepted until far more recently. Within the past five years, a series of studies has shown that stem cells occur in the adult mammalian brain and that these cells can generate its three major cell lineages.

Today, scientists believe that stem cells in the fetal and adult brain divide and give rise to more stem cells or to several types of precursor cells. Neuronal precursors (also called neuroblasts) divide and give rise to nerve cells (neurons), of which there are many types. Glial precursors give rise to astrocytes or oligodendrocytes. Astrocytes are a kind of glial cell, which lend both mechanical and metabolic support for neurons; they make up 70 to 80 percent of the cells of the adult brain. Oligodendrocytes make myelin, the fatty material that ensheathes nerve cell axons and speeds nerve transmission. Under normal, *in vivo* conditions, neuronal precursors do not give rise to glial cells, and glial precursors do not give rise to neurons. In contrast, a fetal or adult CNS (central nervous system the brain and spinal cord) stem cell may give rise to neurons, astrocytes, or oligodendrocytes, depending on the signals it receives and its three-dimensional environment within the brain tissue.

There is now widespread consensus that the adult mammalian brain does contain stem cells. However, there is no consensus about how many populations of CNS stem cells exist, how they may be related, and how they function *in vivo*. Because there are no markers currently available to identify the cells *in vivo*, the only method for testing whether a given population of CNS cells contains stem cells is to isolate the cells and manipulate them *in vitro*, a process that may change their intrinsic properties.

Despite these barriers, three groups of CNS stem cells have been reported to date. All occur in the adult rodent brain and preliminary evidence indicates they also occur in the adult human brain. One group occupies the brain tissue next to the ventricles, regions known as the ventricular zone and the sub-ventricular zone. The ventricles are spaces in the brain filled with cerebrospinal fluid. During fetal development, the tissue adjacent to the ventricles is a prominent region of actively dividing cells. By adulthood, however, this tissue is much smaller, although it still appears to contain stem cells.

A second group of adult CNS stem cells, described in mice but not in humans, occurs in a streak of tissue that connects the lateral ventricle and the olfactory bulb, which receives odour signals from the nose. In rodents, olfactory bulb neurons are constantly being replenished via this pathway. A third possible location for stem cells in adult mouse and human brain occurs in the hippocampus, a part of the brain thought to play a role in the formation of certain kinds of memory.

Central Nervous System Stem Cells in the Subventricular Zone: CNS stem cells found in the forebrain that surrounds the lateral ventricles are heterogeneous and can be distinguished morphologically. Ependymal cells, which are ciliated, line the ventricles. Adjacent to the ependymal cell layer, in a region sometimes designated as the subependymal or subventricular zone, is a mixed cell population that consists of neuroblasts (immature neurons) that migrate to the olfactory bulb, precursor cells, and astrocytes. Some of the cells divide rapidly, while others divide slowly. The astrocyte-like cells can be identified because they contain glial fibrillary acidic protein (GFAP), whereas the ependymal cells stain positive for nestin, which is regarded as a marker of neural stem cells. Which of these cells best qualifies as a CNS stem cell is a matter of debate.

A recent report indicates that the astrocytes that occur in the subventricular zone of the rodent brain act as neural stem cells. The cells with astrocyte markers appear to generate neurons *in vivo*, as identified by their expression of specific neuronal markers. The *in vitro* assay to demonstrate that these astrocytes are, in fact, stem cells involves their ability to form neurospheres groupings of undifferentiated cells that can be dissociated and coaxed to differentiate into neurons or glial cells. Traditionally, these astrocytes have been regarded as differentiated cells, not as stem cells and so their designation as stem cells is not universally accepted.

A series of similar *in vitro* studies based on the formation of neurospheres was used to identify the subependymal zone as a source of adult rodent CNS stem cells. In these experiments, single, candidate stem cells derived from the subependymal zone are induced to give rise to neurospheres in the presence of mitogens either epidermal growth factor (EGF) or fibroblast growth factor-2 (FGF-2).

The neurospheres are dissociated and passaged. As long as a mitogen is present in the culture medium, the cells continue forming neurospheres without differentiating. Some populations of CNS cells are more responsive to EGF, others to FGF. To induce differentiation into neurons or glia, cells are dissociated from the neurospheres and grown on an adherent surface in serum-free medium that contains specific growth factors. Collectively, the studies demonstrate that a population of cells derived from the adult rodent brain can self-renew and differentiate to yield the three major cell types of the CNS cells.

Central Nervous System Stem Cells in the Ventricular Zone: Another group of potential CNS stem cells in the adult rodent brain may consist of the ependymal cells themselves. Ependymal cells, which are ciliated, line the lateral ventricles. They have been described as non-dividing cells that function as part of the blood-brain barrier. The suggestion that ependymal cells from the ventricular zone of the adult rodent CNS may be stem cells is therefore unexpected. However, in a recent study, in which two molecular tags the fluorescent marker Dil, and an adenovirus vector carrying *lacZ* tags were used to label the ependymal cells that line the entire CNS ventricular system of adult rats, it was shown that these cells could, indeed, act as stem cells. A few days after labelling, fluorescent or *lacZ* cells were observed in the rostral migratory stream (which leads from the lateral ventricle to the olfactory bulb), and then in the olfactory bulb itself.

To show that Dil$^+$ cells were neural stem cells and could generate astrocytes and oligodendrocytes as well as neurons, a neurosphere assay was performed *in vitro*. Dil-labelled cells were dissociated from the ventricular system and cultured in the presence of mitogen to generate neurospheres. Most of the neurospheres were Dil$^+$; they could self-renew and generate neurons, astrocytes, and oligodendrocytes when induced to differentiate. Single, Dil$^+$ ependymal cells isolated from the ventricular zone could also generate self-renewing neurospheres and differentiate into neurons and glia. To show that ependymal cells can also divide *in vivo*, bromodeoxyuridine (BrdU) was administered in the drinking water to rats for a 2 to 6 week period. Bromodeoxyuridine (BrdU) is a DNA precursor that is only incorporated into dividing cells. Through a series of experiments, it was shown that ependymal cells divide slowly *in vivo* and give rise to a population of progenitor cells in the subventricular zone. A different pattern of scattered BrdU-labelled cells was observed in the spinal cord, which suggested that ependymal cells along the central canal of the cord occasionally divide and give rise to nearby ependymal cells, but do not migrate away from the canal.

Collectively, the data suggest that CNS ependymal cells in adult rodents can function as stem cells. The cells can self-renew, and most proliferate via asymmetrical division. Many of the CNS ependymal cells are not actively dividing (quiescent), but they can be stimulated to do so *in vitro* (with mitogens) or *in vivo* (in response to injury). After injury, the ependymal cells in the spinal cord only give rise to astrocytes, not to neurons. How and whether ependymal cells from the ventricular zone are related to other candidate populations of CNS stem cells, such as those identified in the hippocampus, is not known.

Are ventricular and subventricular zone CNS stem cells the same population? These studies and other leave open the question of whether cells that directly line the ventricles those in the ventricular zone or cells that are at least a layer removed from this zone in the subventricular zone are the same population of CNS stem cells. A new study, based on the finding that they express different genes, confirms earlier reports that the ventricular and subventricular zone cell populations are distinct.

The new research utilizes a technique called representational difference analysis, together with cDNA microarray analysis, to monitor the patterns of gene expression in the complex tissue of the developing and postnatal mouse brain. The study revealed the expression of a panel of genes known to be important in CNS development, such as L3-PSP (which encodes a phosphoserine phosphatase important in cell signalling), cyclin D2 (a cell cycle gene), and ERCC-1 (which is important in DNA excision repair). All of these genes in the recent study were expressed in cultured neurospheres, as well as the ventricular zone, the subventricular zone, and a brain area outside those germinal zones.

This analysis also revealed the expression of novel genes such as A16F10, which is similar to a gene in an embryonic cancer cell line. A16F10 was expressed in neurospheres and at high levels in the subventricular zone, but not significantly in the ventricular zone. Interestingly, several of the genes identified in cultured neurospheres were also expressed in hematopoietic cells, suggesting that neural stem cells and blood-forming cells may share aspects of their genetic programs or signalling systems. This finding may help explain recent reports that CNS stem cells derived from mouse brain can give rise to hematopoietic cells after injection into irradiated mice.

Central Nervous System Stem Cells in the Hippocampus: The hippocampus is one of the oldest parts of the cerebral cortex, in evolutionary terms, and is thought to play an important role in certain forms of memory. The region of the hippocampus in which stem cells apparently exist in mouse and human brains is the subgranular zone of the dentate gyrus. In mice, when BrdU is used to label dividing cells in this region, about 50% of the labelled cells differentiate into cells that appear to be dentate gyrus granule neurons, and 15% become glial cells.

The rest of the BrdU-labelled cells do not have a recognizable phenotype. Interestingly, many, if not all the BrdU-labelled cells in the adult rodent hippocampus occur next to blood vessels. In the human dentate gyrus, some BrdU-labelled cells express NeuN, neuron-specific enolase, or calbindin, all of which are neuronal markers.

The labelled neuron-like cells resemble dentate gyrus granule cells, in terms of their morphology (as they did in mice). Other BrdU-labelled cells express glial fibrillary acidic protein (GFAP) an astrocyte marker. The study involved autopsy material, obtained with family consent, from five cancer patients who had been injected with BrdU dissolved in saline prior to their death for diagnostic purposes. The patients ranged in age from 57 to 72 years. The greatest number of BrdU-labelled cells were identified in the oldest patient, suggesting that new neuron formation in the hippocampus can continue late in life.

Fetal Central Nervous System Stem Cells: Not surprisingly, fetal stem cells are numerous in fetal tissues, where they are assumed to play an important role in the expansion and differentiation of all tissues of the developing organism. Depending on the developmental stage of an animal, fetal stem cells and precursor cells which arise from stem cells may make up the bulk of a tissue. This is certainly true in the brain, although it has not been demonstrated experimentally in many tissues.

It may seem obvious that the fetal brain contains stem cells that can generate all the types of neurons in the brain as well as astrocytes and oligodendrocytes, but it was not until fairly recently that the concept was proven experimentally. There has been a long-standing question as to whether or not the same cell type gives rise to both neurons and glia. In studies of the developing rodent brain, it has now been shown that all the major cell types in the fetal brain arise from a common population of progenitor cells.

Neural stem cells in the mammalian fetal brain are concentrated in seven major areas: olfactory bulb, ependymal (ventricular) zone of the lateral ventricles (which lie in the forebrain), subventricular zone (next to the ependymal zone), hippocampus, spinal cord, cerebellum (part of the hindbrain), and the cerebral cortex.

Their number and pattern of development vary in different species. These cells appear to represent different stem cell populations, rather than a single population of stem cells that is dispersed in multiple sites. The normal development of the brain depends not only on the proliferation and differentiation of these fetal stem cells, but also on a genetically programmed process of selective cell death called apoptosis.

Little is known about stem cells in the human fetal brain. In one study, however, investigators derived clonal cell lines from CNS stem cells isolated from the diencephalon and cortex of human fetuses, 10.5 weeks post-conception. The study is unusual, not only because it involves human CNS stem cells obtained from fetal tissue, but also because the cells were used to generate clonal cell lines of CNS stem cells that generated neurons, astrocytes, and oligodendrocytes, as determined on the basis of expressed markers. In a few experiments described as "preliminary," the human CNS stem cells were injected into the brains of immunosuppressed rats where they apparently differentiated into neuron-like cells or glial cells.

In a 1999 study, a serum-free growth medium that included EGF and FGF2 was devised to grow the human fetal CNS stem cells. Although most of the cells died, occasionally, single CNS stem cells survived, divided, and ultimately formed neurospheres after one to two weeks in culture. The neurospheres could be dissociated and individual cells replated.

The cells resumed proliferation and formed new neurospheres, thus establishing an *in vitro* system that (like the system established for mouse CNS neurospheres) could be maintained up to 2 years. Depending on the culture conditions, the cells in the neurospheres could be maintained in an undifferentiated dividing state (in the presence of mitogen), or dissociated and induced to differentiate (after the removal of mitogen and the addition of specific growth factors to the culture medium).

The differentiated cells consisted mostly of astrocytes (75%), some neurons (13%) and rare oligodendrocytes (1.2%). The neurons generated under these conditions expressed markers indicating they were GABAergic, [the major type of inhibitory neuron in the mammalian CNS responsive to the amino acid neurotransmitter, gammaaminobutyric acid (GABA)]. However, catecholamine-like cells that express tyrosine hydroxylase (TH, a critical enzyme in the dopamine-synthesis pathway) could be generated, if the culture conditions were altered to include different medium conditioned by a rat glioma line (BB49). Thus, the report indicates that human CNS stem cells obtained from early fetuses can be maintained *in vitro* for a long time without differentiating, induced to differentiate into the three major lineages of the CNS (and possibly two kinds of neurons, GABAergic and TH-positive), and engraft (in rats) in vivo.

Central Nervous System Neural Crest Stem Cells: Neural crest cells differ markedly from fetal or adult neural stem cells. During fetal development, neural crest cells migrate from the sides of the neural tube as it closes. The cells differentiate into a range of tissues, not all of which are part of the nervous system. Neural crest cells form the sympathetic and parasympathetic components of the peripheral nervous system (PNS), including the network of nerves that innervate the heart and the gut, all the sensory ganglia (groups of neurons that occur in pairs along the dorsal surface of the spinal cord), and Schwann cells, which (like oligodendrocytes in the CNS) make myelin in the PNS. The non-neural tissues that arise from the neural crest are diverse. They populate certain hormone-secreting glands including the adrenal medulla and Type I cells in the carotid body pigment cells of the skin (melanocytes), cartilage and bone in the face and skull, and connective tissue in many parts of the body.

Thus, neural crest cells migrate far more extensively than other fetal neural stem cells during development, form mesenchymal tissues, most of which develop from embryonic mesoderm as well as the components of the CNS and PNS which arises from embryonic ectoderm. This close link, in neural crest development, between ectodermally derived tissues and mesodermally derived tissues accounts in part for the interest in neural crest cells as a kind of stem cell. In fact, neural crest cells meet several criteria of stem cells. They can self-renew (at least in the fetus) and can differentiate into multiple cells types, which include cells derived from two of the three embryonic germ layers.

Recent studies indicate that neural crest cells persist late into gestation and can be isolated from E14.5 rat sciatic nerve, a peripheral nerve in the hindlimb. The cells incorporate BrdU, indicating that they are dividing *in vivo*. When transplanted into chick embryos, the rat neural crest cells develop into neurons and glia, an indication of their stem cell-like properties.

However, the ability of rat E14.5 neural crest cells taken from sciatic nerve to generate nerve and glial cells in chick is more limited than neural crest cells derived from younger, E10.5 rat embryos. At the earlier stage of development, the neural tube has formed, but neural crest cells have not yet migrated to their final destinations. Neural crest cells from early developmental stages are more sensitive to bone morphogenetic protein 2 (BMP2) signalling, which may help explain their greater differentiation potential.

Stem Cells in the Bone Marrow and Blood

The notion that the bone marrow contains stem cells is not new. One population of bone marrow cells, the hematopoietic stem cells (HSCs), is responsible for forming all of the types of blood cells in the body. HSCs were recognised as a stem cells more than 40 years ago. Bone marrow stromal cells a mixed cell population that generates bone, cartilage, fat, fibrous connective tissue, and the reticular network that supports blood cell formation were described shortly after the discovery of HSCs. The mesenchymal stem cells of the bone marrow also give rise to these tissues, and may constitute the same population of cells as the bone marrow stromal cells.

Recently, a population of progenitor cells that differentiates into endothelial cells, a type of cell that lines the blood vessels, was isolated from circulating blood and identified as originating in bone marrow. Whether these endothelial progenitor cells, which resemble the angioblasts that give rise to blood vessels during embryonic development, represent a bona fide population of adult bone marrow stem cells remains uncertain. Thus, the bone marrow appears to contain three stem cell populations hematopoietic stem cells, stromal cells, and (possibly) endothelial progenitor cells

Two more apparent stem cell types have been reported in circulating blood, but have not been shown to originate from the bone marrow. One population, called pericytes, may be closely related to bone marrow stromal cells, although their origin remains elusive. The second population of blood-born stem cells, which occur in four species of animals tested guinea pigs, mice, rabbits, and humans resemble stromal cells in that they can generate bone and fat.

Hematopoietic Stem Cells

Of all the cell types in the body, those that survive for the shortest period of time are blood cells and certain kinds of epithelial cells. For example, red blood cells (erythrocytes), which lack a nucleus, live for approximately 120 days in the bloodstream. The life of an animal literally depends on the ability of these and other blood cells to be replenished continuously. This replenishment process occurs largely in the bone marrow, where HSCs reside, divide, and differentiate into all the blood cell types. Both HSCs and differentiated blood cells cycle from the bone marrow to the blood and back again, under the influence of a barrage of secreted factors that regulate cell proliferation, differentiation, and migration.

HSCs can reconstitute the hematopoietic system of mice that have been subjected to lethal doses of radiation to destroy their own hematopoietic systems. This test, the rescue of lethally irradiated mice, has become a standard by which other candidate stem cells are measured because it shows, without question, that HSCs can regenerate an entire tissue system in this case, the blood. HSCs were first proven to be blood-forming stem cells in a series of experiments in mice; similar blood-forming stem cells occur in humans. HSCs are defined by their ability to self-renew and to give rise to all the kinds of blood cells in the body. This means that a single HSC is capable of regenerating the entire hematopoietic system, although this has been demonstrated only a few times in mice.

Over the years, many combinations of surface markers have been used to identify, isolate, and purify HSCs derived from bone marrow and blood. Undifferentiated HSCs and hematopoietic progenitor cells express c-kit, CD34, and H-2K. These cells usually lack the lineage marker Lin, or express it at very low levels ($Lin^{-/low}$). And for transplant purposes, cells that are $CD34^+$ $Thy1^+$ Lin^- are most likely to contain stem cells and result in engraftment.

Two kinds of HSCs have been defined. Long-term HSCs proliferate for the lifetime of an animal. In young adult mice, an estimated 8 to 10 % of long-term HSCs enter the cell cycle and divide each day. Short-term HSCs proliferate for a limited time, possibly a few months. Long-term HSCs have high levels of telomerase activity. Telomerase is an enzyme that helps maintain the length of the ends of chromosomes, called telomeres, by adding on nucleotides. Active telomerase is a characteristic of undifferentiated, dividing cells and cancer cells. Differentiated, human somatic cells do not show telomerase activity. In adult humans, HSCs occur in the bone marrow, blood, liver, and spleen, but are extremely rare in any of these tissues. In mice, only 1 in 10,000 to 15,000 bone marrow cells is a long-term HSC.

Short-term HSCs differentiate into lymphoid and myeloid precursors, the two classes of precursors for the two major lineages of blood cells. Lymphoid precursors differentiate into T cells, B cells, and natural killer cells. The mechanisms and pathways that lead to their differentiation are still being investigated. Myeloid precursors differentiate into monocytes and macrophages, neutrophils, eosinophils, basophils, megakaryocytes, and erythrocytes. *In vivo*, bone marrow HSCs differentiate into mature, specialised blood cells that cycle constantly from the bone marrow to the blood, and back to the bone marrow. A recent study showed that short-term HSCs are a heterogeneous population that differ significantly in terms of their ability to self-renew and repopulate the hematopoietic system.

Attempts to induce HSC to proliferate *in vitro* on many substrates, including those intended to mimic conditions in the stroma have frustrated scientists for many years. Although HSCs proliferate readily *in vivo*, they usually differentiate or die *in vitro*. Thus, much of the research on HSCs has been focused on understanding the factors, cell-cell interactions, and cell-matrix interactions that control their proliferation and differentiation *in vivo*, with the hope that similar conditions could be replicated *in vitro*. Many of the soluble factors that regulate HSC differentiation *in vivo* are cytokines, which are made by different cell types and are then concentrated in the bone marrow by the extracellular matrix of stromal cells the sites of blood formation. Two of the most-studied cytokines are granulocyte-macrophage colony-stimulating factor (GM-CSF) and interleukin-3 (IL-3).

Also important to HSC proliferation and differentiation are interactions of the cells with adhesion molecules in the extracellular matrix of the bone marrow stroma.

Bone Marrow Stromal Cells

Bone marrow (BM) stromal cells have long been recognised for playing an important role in the differentiation of mature blood cells from HSCs. But stromal cells also have other important functions. In addition to providing the physical environment in which HSCs differentiate, BM stromal cells generate cartilage, bone, and fat. Whether stromal cells are best classified as stem cells or progenitor cells for these tissues is still in question. There is also a question as to whether BM stromal cells and so-called mesenchymal stem cells are the same population.

BM stromal cells have many features that distinguish them from HSCs. The two cell types are easy to separate *in vitro*. When bone marrow is dissociated, and the mixture of cells it contains is plated at low density, the stromal cells adhere to the surface of the culture dish, and the HSCs do not. Given specific *in vitro* conditions, BM stromal cells form colonies from a single cell called the colony forming unit-F (CFU-F). These colonies may then differentiate as adipocytes or myelosupportive stroma, a clonal assay that indicates the stem cell-like nature of stromal cells. Unlike HSCs, which do not divide *in vitro* (or proliferate only to a limited extent), BM stromal cells can proliferate for up to 35 population doublings *in vitro*. They grow rapidly under the influence of such mitogens as platelet-derived growth factor (PDGF), epidermal growth factor (EGF), basic fibroblast growth factor (bFGF), and insulin-like growth factor-1 (IGF-1).

To date, it has not been possible to isolate a population of pure stromal cells from bone marrow. Panels of markers used to identify the cells include receptors for certain cytokines (interleukin-1, 3, 4, 6, and 7) receptors for proteins in the extracellular matrix, (ICAM-1 and 2, VCAM-1, the alpha-1, 2, and 3 integrins, and the beta-1, 2, 3 and 4 integrins), etc. Despite the use of these markers and another stromal cell marker called Stro-1, the origin and specific identity of stromal cells have remained elusive. Like HSCs, BM stromal cells arise from embryonic mesoderm during development, although no specific precursor or stem cell for stromal cells has been isolated and identified. One theory about their origin is that a common kind of progenitor cell perhaps a primordial endothelial cell that lines embryonic blood vessels gives rise to both HSCs and to mesodermal precursors. The latter may then differentiate into myogenic precursors (the satellite cells that are thought to function as stem cells in skeletal muscle), and the BM stromal cells.

In vivo, the differentiation of stromal cells into fat and bone is not straightforward. Bone marrow adipocytes and myelosupportive stromal cells both of which are derived from BM stromal cells may be regarded as interchangeable phenotypes. Adipocytes do not develop until postnatal life, as the bones enlarge and the marrow space increases to accommodate enhanced hematopoiesis. When the skeleton stops growing, and the mass of HSCs decreases in a normal, age-dependent fashion, BM stromal cells differentiate into adipocytes, which fill the extra space.

New bone formation is obviously greater during skeletal growth, although bone "turns over" throughout life. Bone forming cells are osteoblasts, but their relationship to BM stromal cells is not clear. New trabecular bone, which is the inner region of bone next to the marrow, could logically develop from the action of BM stromal cells. But the outside surface of bone also turns over, as does bone next to the Haversian system (small canals that form concentric rings within bone). And neither of these surfaces is in contact with BM stromal cells.

Adult Stem Cells in Other Tissues

It is often difficult if not impossible to distinguish adult, tissue-specific stem cells from progenitor cells. With that caveat in mind, the following summary identifies reports of stem cells in various adult tissues.

Endothelial Progenitor Cells

Endothelial cells line the inner surfaces of blood vessels throughout the body, and it has been difficult to identify specific endothelial stem cells in either the embryonic or the adult mammal. During embryonic development, just after gastrulation, a kind of cell called the hemangioblast, which is derived from mesoderm, is presumed to be the precursor of both the hematopoietic and endothelial cell lineages. The embryonic vasculature formed at this stage is transient and consists of blood islands in the yolk sac. But hemangioblasts, per se, have not been isolated from the embryo and their existence remains in question. The process of forming new blood vessels in the embryo is called vasculogenesis. In the adult, the process of forming blood vessels from pre-existing blood vessels is called angiogenesis.

Evidence that hemangioblasts do exist comes from studies of mouse embryonic stem cells that are directed to differentiate *in vitro*. These studies have shown that a precursor cell derived from mouse ES cells that express Flk-1 [the receptor for vascular endothelial growth factor (VEGF) in mice] can give rise to both blood cells and blood vessel cells. Both VEGF and fibroblast growth factor-2 (FGF-2) play critical roles in endothelial cell differentiation *in vivo*.

Several recent reports indicate that the bone marrow contains cells that can give rise to new blood vessels in tissues that are ischemic (damaged due to the deprivation of blood and oxygen). But it is unclear from these studies what cell type(s) in the bone marrow induced angiogenesis. In a study which sought to address that question, researchers found that adult human bone marrow contains cells that resemble embryonic hemangioblasts, and may therefore be called endothelial stem cells.

In more recent experiments, human bone marrow-derived cells were injected into the tail veins of rats with induced cardiac ischemia. The human cells migrated to the rat heart where they generated new blood vessels in the infarcted muscle (a process akin to vasculogenesis), and also induced angiogenesis. The candidate endothelial stem cells are $CD34^+$(a marker for HSCs), and they express the transcription factor GATA-2. A similar study using transgenic mice that express the gene for enhanced green fluorescent protein (which allows the cells to be tracked), showed that bone-marrow-derived cells could repopulate an area of infarcted heart muscle in mice, and generate not only blood vessels, but also cardiomyocytes that integrated into the host tissue.

And, in a series of experiments in adult mammals, progenitor endothelial cells were isolated from peripheral blood (of mice and humans) by using antibodies against CD34 and Flk-1, the receptor for VEGF. The cells were mononuclear blood cells (meaning they have a nucleus) and are referred to as MB^{CD34+} cells and MB^{Flk1+} cells. When plated in tissue-culture dishes, the cells attached to the substrate, became spindle-shaped, and formed tube-like structures that resemble blood vessels. When transplanted into mice of the same species (autologous transplants) with induced ischemia in one limb, the MB^{CD34+} cells promoted the formation of new blood vessels. Although the adult MB^{CD34+} and MB^{Flk1+} cells function in some ways like stem cells, they are usually regarded as progenitor cells.

Skeletal Muscle Stem Cells

Skeletal muscle, like the cardiac muscle of the heart and the smooth muscle in the walls of blood vessels, the digestive system, and the respiratory system, is derived from embryonic mesoderm. To date, at least three populations of skeletal muscle stem cells have been identified: satellite cells, cells in the wall of the dorsal aorta, and so-called "side population" cells. Satellite cells in skeletal muscle were identified 40 years ago in frogs by electron microscopy, and thereafter in mammals.

Satellite cells occur on the surface of the basal lamina of a mature muscle cell, or myofibre. In adult mammals, satellite cells mediate muscle growth. Although satellite cells are normally non-dividing, they can be triggered to proliferate as a result of injury, or weight-bearing exercise. Under either of these circumstances, muscle satellite cells give rise to myogenic precursor cells, which then differentiate into the myofibrils that typify skeletal muscle. A group of transcription factors called myogenic regulatory factors (MRFs) play important roles in these differentiation events. The so-called primary MRFs, MyoD and Myf5, help regulate myoblast formation during embryogenesis. The secondary MRFs, myogenin and MRF4, regulate the terminal differentiation of myofibrils.

With regard to satellite cells, scientists have been addressing two questions. Are skeletal muscle satellite cells true adult stem cells or are they instead precursor cells? Are satellite cells the only cell type that can regenerate skeletal muscle. For example, a recent report indicates that muscle stem cells may also occur in the dorsal aorta of mouse embryos, and constitute a cell type that gives rise both to muscle satellite cells and endothelial cells. Whether the dorsal aorta cells meet the criteria of a self-renewing muscle stem cell is a matter of debate.

Another report indicates that a different kind of stem cell, called an SP cell, can also regenerate skeletal muscle may be present in muscle and bone marrow. SP stands for a side population of cells that can be separated by fluorescence-activated cell sorting analysis.

Intravenously injecting these muscle-derived stem cells restored the expression of dystrophin in mdx mice. Dystrophin is the protein that is defective in people with Duchenne's muscular dystrophy; mdx mice provide a model for the human disease. Dystrophin expression in the SP cell-treated mice was lower than would be needed for clinical benefit. Injection of bone marrow- or muscle-derived SP cells into the dystrophic muscle of the mice yielded equivocal results that the transplanted cells had integrated into the host tissue.

The authors conclude that a similar population of SP stem cells can be derived from either adult mouse bone marrow or skeletal muscle, and suggest "there may be some direct relationship between bone marrow-derived stem cells and other tissue or organ-specific cells". Thus, stem cell or progenitor cell types from various mesodermally-derived tissues may be able to generate skeletal muscle.

Epithelial Cell Precursors in the Skin and Digestive System

Epithelial cells, which constitute 60 percent of the differentiated cells in the body are responsible for covering the internal and external surfaces of the body, including the lining of vessels and other cavities. The epithelial cells in skin and the digestive tract are replaced constantly. Other epithelial cell populations in the ducts of the liver or pancreas, for example turn over more slowly. The cell population that renews the epithelium of the small intestine occurs in the intestinal crypts, deep invaginations in the lining of the gut. The crypt cells are often regarded as stem cells; one of them can give rise to an organised cluster of cells called a structural-proliferative unit.

The skin of mammals contains at least three populations of epithelial cells: epidermal cells, hair follicle cells, and glandular epithelial cells, such as those that make up the sweat glands. The replacement patterns for epithelial cells in these three compartments differ, and in all the compartments, a stem cell population has been postulated. For example, stem cells in the bulge region of the hair follicle appear to give rise to multiple cell types. Their progeny can migrate down to the base of the follicle where they become matrix cells, which may then give rise to different cell types in the hair follicle, of which there are seven. The bulge stem cells of the follicle may also give rise to the epidermis of the skin.

Another population of stem cells in skin occurs in the basal layer of the epidermis. These stem cells proliferate in the basal region, and then differentiate as they move toward the outer surface of the skin. The keratinocytes in the outermost layer lack nuclei and act as a protective barrier. A dividing skin stem cell can divide asymmetrically to produce two kinds of daughter cells. One is another self-renewing stem cell. The second kind of daughter cell is an intermediate precursor cell which is then committed to replicate a few times before differentiating into keratinocytes. Self-renewing stem cells can be distinguished from this intermediate precusor cell by their higher level of â1 integrin expression, which signals keratinocytes to proliferate via a mitogen-activated protein (MAP) kinase. Other signalling pathways include that triggered by catenin, which helps maintain the stem-cell state, and the pathway regulated by the oncoprotein c-Myc, which triggers stem cells to give rise to transit mplifying cells.

Stem Cells in the Pancreas and Liver

The status of stem cells in the adult pancreas and liver is unclear. During embryonic development, both tissues arise from endoderm. A recent study indicates that a single precursor cell derived from embryonic endoderm may generate both the ventral pancreas and the liver. In adult mammals, however, both the pancreas and the liver contain multiple kinds of differentiated cells that may be repopulated or regenerated by multiple types of stem cells. In the pancreas, endocrine (hormone-producing) cells occur in the islets of Langerhans. They include the beta cells (which produce insulin), the alpha cells (which secrete glucagon), and cells that release the peptide hormones somatostatin and pancreatic polypeptide. Stem cells in the adult pancreas are postulated to occur in the pancreatic ducts or in the islets themselves. Several recent reports indicate that stem cells that express nestin which is usually regarded as a marker of neural stem cells can generate all of the cell types in the islets.

The identity of stem cells that can repopulate the liver of adult mammals is also in question. Recent studies in rodents indicate that HSCs (derived from mesoderm) may be able to home to liver after it is damaged, and demonstrate plasticity in becoming into hepatocytes (usually derived from endoderm). But the question remains as to whether cells from the bone marrow normally generate hepatocytes *in vivo*.

It is not known whether this kind of plasticity occurs without severe damage to the liver or whether HSCs from the bone marrow generate oval cells of the liver. Although hepatic oval cells exist in the liver, it is not clear whether they actually generate new hepatocytes. Oval cells may arise from the portal tracts in liver and may give rise to either hepatocytes and to the epithelium of the bile ducts. Indeed, hepatocytes themselves, may be responsible for the well-know regenerative capacity of liver.

Stem Cells and Diabetes

Diabetes exacts its toll on many Americans, young and old. For years, researchers have painstakingly dissected this complicated disease caused by the destruction of insulin producing islet cells of the pancreas. Despite progress in understanding the underlying disease mechanisms for diabetes, there is still a paucity of effective therapies. For years investigators have been making slow, but steady, progress on experimental strategies for pancreatic transplantation and islet cell replacement. Now, researchers have turned their attention to adult stem cells that appear to be precursors to islet cells and embryonic stem cells that produce insulin.

For decades, diabetes researchers have been searching for ways to replace the insulin-producing cells of the pancreas that are destroyed by a patient's own immune system. Now it appears that this may be possible. Each year, diabetes affects more people and causes more deaths than breast cancer and AIDS combined. Diabetes is the seventh leading cause of death in the United States today, with nearly 200,000 deaths reported each year. The American Diabetes Association estimates that nearly 16 million people, or 5.9 percent of the United States population, currently have diabetes.

Diabetes is actually a group of diseases characterised by abnormally high levels of the sugar glucose in the bloodstream. This excess glucose is responsible for most of the complications of diabetes, which include blindness, kidney failure, heart disease, stroke, neuropathy, and amputations. Type 1 diabetes, also known as juvenile-onset diabetes, typically affects children and young adults. Diabetes develops when the body's immune system sees its own cells as foreign and attacks and destroys them.

As a result, the islet cells of the pancreas, which normally produce insulin, are destroyed. In the absence of insulin, glucose cannot enter the cell and glucose accumulates in the blood. Type 2 diabetes, also called adult-onset diabetes, tends to affect older, sedentary, and overweight individuals with a family history of diabetes. Type 2 diabetes occurs when the body cannot use insulin effectively. This is called insulin resistance and the result is the same as with type 1 diabetes a build up of glucose in the blood.

There is currently no cure for diabetes. People with type 1 diabetes must take insulin several times a day and test their blood glucose concentration three to four times a day throughout their entire lives. Frequent monitoring is important because patients who keep their blood glucose concentrations as close to normal as possible can significantly reduce many of the complications of diabetes, such as retinopathy (a disease of the small blood vessels of the eye which can lead to blindness) and heart disease, that tend to develop over time. People with type 2 diabetes can often control their blood glucose concentrations through a combination of diet, exercise, and oral medication. Type 2 diabetes often progresses to the point where only insulin therapy will control blood glucose concentrations.

Each year, approximately 1,300 people with type 1 diabetes receive whole-organ pancreas transplants. After a year, 83 percent of these patients, on average, have no symptoms of diabetes and do not have to take insulin to maintain normal glucose concentrations in the blood. However, the demand for transplantable pancreases outweighs their availability. To prevent the body from rejecting the transplanted pancreas, patients must take powerful drugs that suppress the immune system for their entire lives, a regimen that makes them susceptible to a host of other diseases. Many hospitals will not perform a pancreas transplant unless the patient also needs a kidney transplant. That is because the risk of infection due to immunosuppressant therapy can be a greater health threat than the diabetes itself. But if a patient is also receiving a new kidney and will require immunosuppressant drugs anyway, many hospitals will perform the pancreas transplant.

Over the past several years, doctors have attempted to cure diabetes by injecting patients with pancreatic islet cells the cells of the pancreas that secrete insulin and other hormones. However, the requirement for steroid immunosuppressant therapy to prevent rejection of the cells increases the metabolic demand on insulin-producing cells and eventually they may exhaust their capacity to produce insulin. The deleterious effect of steroids is greater for islet cell transplants than for whole-organ transplants. As a result, less than 8 percent of islet cell transplants performed before last year had been successful.

More recently, James Shapiro and his colleagues in Edmonton, Alberta, Canada, have developed an experimental protocol for transplanting islet cells that involves using a much larger amount of islet cells and a different type of immunosuppressant therapy. In a recent study, they report that, seven of seven patients who received islet cell transplants no longer needed to take insulin, and their blood glucose concentrations were normal a year after surgery. The success of the Edmonton protocol is now being tested at 10 centres around the world. If the success of the Edmonton protocol can be duplicated, many hurdles still remain in using this approach on a wide scale to treat diabetes. First, donor tissue is not readily available. Islet cells used in transplants are obtained from cadavers, and the procedure requires at least two cadavers per transplant. The islet cells must be immunologically compatible, and the tissue must be freshly obtained within eight hours of death. Because of the shortage of organ donors, these requirements are difficult to meet and the waiting list is expected to far exceed available tissue, especially if the procedure becomes widely accepted and available. Further, islet cell transplant recipients face a lifetime of immunosuppressant therapy, which makes them susceptible to other serious infections and diseases.

Development of the Pancreas

Before discussing cell-based therapies for diabetes, it is important to understand how the pancreas develops. In mammals, the pancreas contains three classes of cell types: the ductal cells, the acinar cells, and the endocrine cells. The endocrine cells produce the hormones glucagon, somatostatin, pancreatic polypeptide (PP), and insulin, which are secreted into the blood stream and help the body regulate sugar metabolism. The acinar cells are part of the exocrine system, which manufactures digestive enzymes, and ductal cells from the pancreatic ducts, which connect the acinar cells to digestive organs.

In humans, the pancreas develops as an outgrowth of the duodenum, a part of the small intestine. The cells of both the exocrine system the acinar cells and of the endocrine system the islet cells seem to originate from the ductal cells during development. During development these endocrine cells emerge from the pancreatic ducts and form aggregates that eventually form what is known as Islets of Langerhans. In humans, there are four types of islet cells: the insulin-producing beta cells; the alpha cells, which produce glucagon; the delta cells, which secrete somatostatin; and the PP-cells, which produce pancreatic polypeptide. The hormones released from each type of islet cell have a role in regulating hormones released from other islet cells. In the human pancreas, 65 to 90 percent of islet cells are beta cells, 15 to 20 percent are alpha-cells, 3 to 10 percent are delta cells, and one percent is PP cells. Acinar cells form small lobules contiguous with the ducts. The resulting pancreas is a combination of a lobulated, branched acinar gland that forms the exocrine pancreas, and, embedded in the acinar gland, the Islets of Langerhans, which constitute the endocrine pancreas.

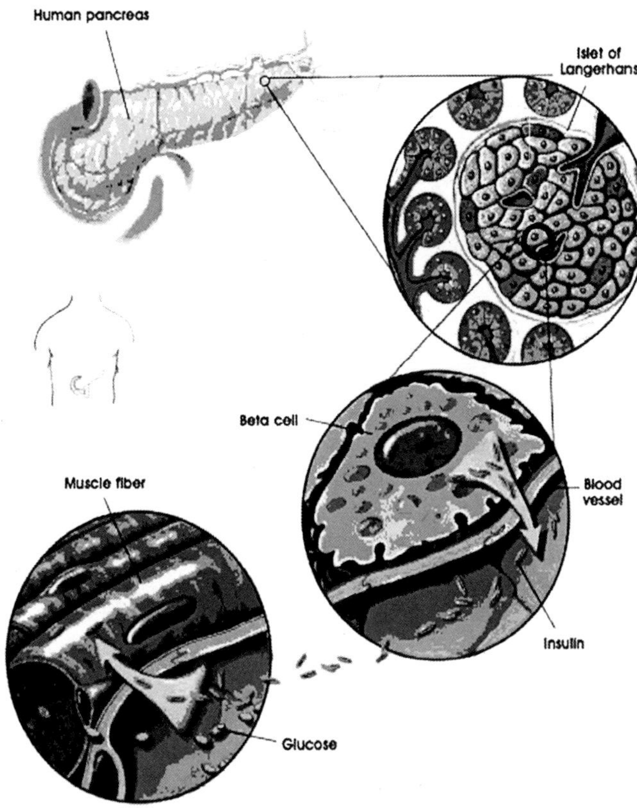

Figure: *Insulin Production in the Human Pancreas. The pancreas is located in the abdomen, adjacent to the duodenum (the first portion of the small intestine). A cross-section of the pancreas shows the islet of Langerhans which is the functional unit of the endocrine pancreas. Encircled is the beta cell that synthesizes and secretes insulin. Beta cells are located adjacent to blood vessels and can easily respond to changes in blood glucose concentration by adjusting insulin production. Insulin facilitates uptake of glucose, the main fuel source, into cells of tissues such as muscle.*

During fetal development, new endocrine cells appear to arise from progenitor cells in the pancreatic ducts. Many researchers maintain that some sort of islet stem cell can be found intermingled with ductal cells during fetal development and that these stem cells give rise to new endocrine cells as the fetus develops. Ductal cells can be distinguished from endocrine cells by their structure and by the genes they express. For example, ductal cells typically express a gene known as *cytokeratin-9 (CK-9),* which encodes a structural protein. Beta islet cells, on the other hand, express a gene called *PDX-1,* which encodes a protein that initiates transcription from the insulin gene. These genes, called cell markers, are useful in identifying particular cell types.

Following birth and into adulthood, the source of new islet cells is not clear, and some controversy exists over whether adult stem cells exist in the pancreas. Some researchers believe that islet stem cell-like cells can be found in the pancreatic ducts and even in the islets themselves. Others maintain that the ductal cells can differentiate into islet precursor cells, while others hold that new islet cells arise from stem cells in the blood. Researchers are using several approaches for isolating and cultivating stem cells or islet precursor cells from fetal and adult pancreatic tissue. In addition, several new promising studies indicate that insulin-producing cells can be cultivated from embryonic stem cell lines.

Development of Cell-Based Therapies for Diabete

In developing a potential therapy for patients with diabetes, researchers hope to develop a system that meets several criteria. Ideally, stem cells should be able to multiply in culture and reproduce themselves exactly. That is, the cells should be self-renewing. Stem cells should also be able to differentiate *in vivo* to produce the desired kind of cell.

For diabetes therapy, it is not clear whether it will be desirable to produce only beta cells the islet cells that manufacture insulin or whether other types of pancreatic islet cells are also necessary. Studies by Bernat Soria and colleagues, for example, indicate that isolated beta cells those cultured in the absence of the other types of islet cells are less responsive to changes in glucose concentration than intact islet clusters made up of all islet cell types. Islet cell clusters typically respond to higher than normal concentrations of glucose by releasing insulin in two phases: a quick release of high concentrations of insulin and a slower release of lower concentrations of insulin.

In this manner the beta cells can fine-tune their response to glucose. Extremely high concentrations of glucose may require that more insulin be released quickly, while intermediate concentrations of glucose can be handled by a balance of quickly and slowly released insulin.

Isolated beta cells, as well as islet clusters with lower than normal amounts of non-beta cells, do not release insulin in this biphasic manner. Instead insulin is released in an all-or-nothing manner, with no fine-tuning for intermediate concentrations of glucose in the blood. Therefore, many researchers believe that it will be preferable to develop a system in which stem or precursor cell types can be cultured to produce all the cells of the islet cluster in order to generate a population of cells that will be able to coordinate the release of the appropriate amount of insulin to the physiologically relevant concentrations of glucose in the blood.

Fetal Tissue as Source for Islet Cells

Several groups of researchers are investigating the use of fetal tissue as a potential source of islet progenitor cells. For example, using mice, researchers have compared the insulin content of implants from several sources of stem cells fresh human fetal pancreatic tissue, purified human islets, and cultured islet tissue. They found that insulin content was initially higher in the fresh tissue and purified islets. However, with time, insulin concentration decreased in the whole tissue grafts, while it remained the same in the purified islet grafts. When cultured islets were implanted, however, their insulin content increased over the course of three months. The researchers concluded that precursor cells within the cultured islets were able to proliferate (continue to replicate) and differentiate (specialize) into functioning islet tissue, but that the purified islet cells (already differentiated) could not further proliferate when grafted. Importantly, the researchers found, however, that it was also difficult to expand cultures of fetal islet progenitor cells in culture.

Adult Tissue as Source for Islet Cells

Many researchers have focused on culturing islet cells from human adult cadavers for use in developing transplantable material. Although differentiated beta cells are difficult to proliferate and culture, some researchers have had success in engineering such cells to do this. For example, Fred Levine and his colleagues at the University of California, San Diego, have engineered islet cells isolated from human cadavers by adding to the cells' DNA special genes that stimulate cell proliferation. However, because once such cell lines that can proliferate in culture are established, they no longer produce insulin. The cell lines are further engineered to express the beta islet cell gene, PDX-1, which stimulates the expression of the insulin gene.

Such cell lines have been shown to propagate in culture and can be induced to differentiate to cells, which produce insulin. When transplanted into immune-deficient mice, the cells secrete insulin in response to glucose. The researchers are currently investigating whether these cells will reverse diabetes in an experimental diabetes model in mice.

These investigators report that these cells do not produce as much insulin as normal islets, but it is within an order of magnitude. The major problem in dealing with these cells is maintaining the delicate balance between growth and differentiation. Cells that proliferate well do not produce insulin efficiently, and those that do produce insulin do not proliferate well. According to the researchers, the major issue is developing the technology to be able to grow large numbers of these cells that will reproducibly produce normal amounts of insulin.

Another promising source of islet progenitor cells lies in the cells that line the pancreatic ducts. Some researchers believe that multipotent (capable of forming cells from more than one germ layer) stem cells are intermingled with mature, differentiated duct cells, while others believe that the duct cells themselves can undergo a differentiation, or a reversal to a less mature type of cell, which can then differentiate into an insulin-producing islet cell.

Susan Bonner-Weir and her colleagues reported last year that when ductal cells isolated from adult human pancreatic tissue were cultured, they could be induced to differentiate into clusters that contained both ductal and endocrine cells. Over the course of three to four weeks in culture, the cells secreted low amounts of insulin when exposed to low concentrations of glucose, and higher amounts of insulin when exposed to higher glucose concentrations. The researchers have determined by immunochemistry and ultrastructural analysis that these clusters contain all of the endocrine cells of the islet.

Bonner-Weir and her colleagues are working with primary cell cultures from duct cells and have not established cells lines that can grow indefinitely. However the cells can be expanded. According to the researchers, it might be possible in principle to do a biopsy and remove duct cells from a patient and then proliferate the cells in culture and give the patient back his or her own islets. This would work with patients who have type 1 diabetes and who lack functioning beta cells, but their duct cells remain intact. However, the autoimmune destruction would still be a problem and potentially lead to destruction of these transplanted cells.

Type 2 diabetes patients might benefit from the transplantation of cells expanded from their own duct cells since they would not need any immunosuppression. However, many researchers believe that if there is a genetic component to the death of beta cells, then beta cells derived from ductal cells of the same individual would also be susceptible to autoimmune attack.

Some researchers question whether the ductal cells are indeed undergoing a dedifferentiation or whether a subset of stem-like or islet progenitors populate the pancreatic ducts and may be co-cultured along with the ductal cells. If ductal cells die off but islet precursors proliferate, it is possible that the islet precursor cells may overtake the ductal cells in culture and make it appear that the ductal cells are dedifferentiating into stem cells. According to Bonner-Weir, both dedifferentiated ductal cells and islet progenitor cells may occur in pancreatic ducts.

Ammon Peck of the University of Florida, Vijayakumar Ramiya of Ixion Biotechnology in Alachua, FL, and their colleagues have also cultured cells from the pancreatic ducts from both humans and mice. Last year, they reported that pancreatic ductal epithelial cells from adult mice could be cultured to yield islet-like structures similar to the cluster of cells found by Bonner-Weir. Using a host of islet-cell markers they identified cells that produced insulin, glucagon, somatostatin, and pancreatic polypeptide. When the cells were implanted into diabetic mice, the diabetes was reversed.

Joel Habener has also looked for islet-like stem cells from adult pancreatic tissue. He and his colleagues have discovered a population of stem-like cells within both the adult pancreas islets and pancreatic ducts. These cells do not express the marker typical of ductal cells, so they are unlikely to be ductal cells, according to Habener. Instead, they express a marker called nestin, which is typically found in developing neural cells. The nestin-positive cells do not express markers typically found in mature islet cells. However, depending upon the growth factors added, the cells can differentiate into different types of cells, including liver, neural, exocrine pancreas, and endocrine pancreas, judged by the markers they express, and can be maintained in culture for up to eight months.

Embryonic Stem Cells

The discovery of methods to isolate and grow human embryonic stem cells in 1998 renewed the hopes of doctors, researchers, and diabetes patients and their families that a cure for type 1 diabetes, and perhaps type 2 diabetes as well, may be within striking distance. In theory, embryonic stem cells could be cultivated and coaxed into developing into the insulin-producing islet cells of the pancreas. With a ready supply of cultured stem cells at hand, the theory is that a line of embryonic stem cells could be grown up as needed for anyone requiring a transplant.

The cells could be engineered to avoid immune rejection. Before transplantation, they could be placed into nonimmunogenic material so that they would not be rejected and the patient would avoid the devastating effects of immunosuppressant drugs. There is also some evidence that differentiated cells derived from embryonic stem cells might be less likely to cause immune rejection. Although having a replenishable supply of insulin-producing cells for transplant into humans may be a long way off, researchers have been making remarkable progress in their quest for it. While some researchers have pursued the research on embryonic stem cells, other researchers have focused on insulin-producing precursor cells that occur naturally in adult and fetal tissues.

Since their discovery three years ago, several teams of researchers have been investigating the possibility that human embryonic stem cells could be developed as a therapy for treating diabetes. Recent studies in mice show that embryonic stem cells can be coaxed into differentiating into insulin-producing beta cells, and new reports indicate that this strategy may be possible using human embryonic cells as well.

Last year, researchers in Spain reported using mouse embryonic stem cells that were engineered to allow researchers to select for cells that were differentiating into insulin-producing cells. Bernat Soria and his colleagues at the Universidad Miguel Hernandez in San Juan, Alicante, Spain, added DNA containing part of the insulin gene to embryonic cells from mice. The insulin gene was linked to another gene that rendered the mice resistant to an antibiotic drug. By growing the cells in the presence of an antibiotic, only those cells that were activating the insulin promoter were able to survive. The cells were cloned and then cultured under varying conditions. Cells cultured in the presence of low concentrations of glucose differentiated and were able to respond to changes in glucose concentration by increasing insulin secretion nearly sevenfold. The researchers then implanted the cells into the spleens of diabetic mice and found that symptoms of diabetes were reversed.

Manfred Ruediger of Cardion, Inc., in Erkrath, Germany, is using the approach developed by Soria and his colleagues to develop insulin-producing human cells derived from embryonic stem cells. By using this method, the non-insulin-producing cells will be killed off and only insulin-producing cells should survive. This is important in ensuring that undifferentiated cells are not implanted that could give rise to tumours. However, some researchers believe that it will be important to engineer systems in which all the components of a functioning pancreatic islet are allowed to develop.

Recently Ron McKay and his colleagues described a series of experiments in which they induced mouse embryonic cells to differentiate into insulin-secreting structures that resembled pancreatic islets. McKay and his colleagues started with embryonic stem cells and let them form embryoid bodies—an aggregate of cells containing all three embryonic germ layers. They then selected a population of cells from the embryoid bodies that expressed the neural marker nestin. Using a sophisticated five-stage culturing technique, the researchers were able to induce the cells to form islet-like clusters that resembled those found in native pancreatic islets. The cells responded to normal glucose concentrations by secreting insulin, although insulin amounts were lower than those secreted by normal islet cells. When the cells were injected into diabetic mice, they survived, although they did not reverse the symptoms of diabetes.

According to McKay, this system is unique in that the embryonic cells form a functioning pancreatic islet, complete with all the major cell types. The cells assemble into islet-like structures that contain another layer, which contains neurons and is similar to intact islets from the pancreas. Several research groups are trying to apply McKay's results with mice to induce human embryonic stem cells to differentiate into insulin-producing islets.

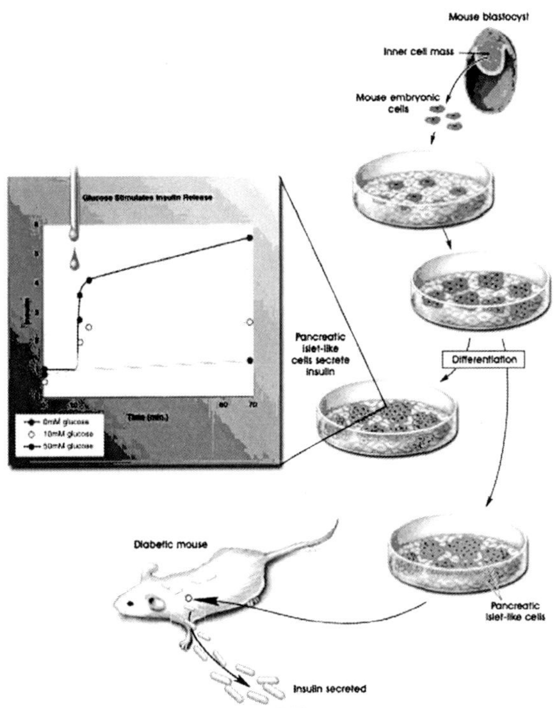

Figure: *Development of Insulin-Secreting Pancreatic-Like Cells From Mouse Embryonic Stem Cells. Mouse embryonic stem cells were derived from the inner cell mass of the early embryo (blastocyst) and cultured under specific conditions. The embryonic stem cells (in blue) were then expanded and differentiated. Cells with markers consistent with islet cells were selected for further differentiation and characterisation. When these cells (in purple) were grown in culture, they spontaneously formed three-dimentional clusters similar in structure to normal pancreatic islets. The cells produced and secreted insulin. As depicted in the chart, the pancreatic islet-like cells showed an increase in release of insulin as the glucose concentration of the culture media was increased. When the pancreatic islet-like cells were implanted in the shoulder of diabetic mice, the cells became vascularised, synthesized insulin, and maintained physical characteristics similar to pancreatic islets.*

Recent research has also provided more evidence that human embryonic cells can develop into cells that can and do produce insulin. Last year, Melton, Nissim Benvinisty of the Hebrew University in Jerusalem, and Josef Itskovitz-Eldor of the Technion in Haifa, Israel, reported that human embryonic stem cells could be manipulated in culture to express the PDX-1 gene, a gene that controls insulin transcription. In these experiments, researchers cultured human embryonic stem cells and allowed them to spontaneously form embryoid bodies (clumps of embryonic stem cells composed of many types of cells from all three germ layers).

The embryoid bodies were then treated with various growth factors, including nerve growth factor. The researchers found that both untreated embryoid bodies and those treated with nerve growth factor expressed PDX-1. Embryonic stem cells prior to formation of the aggregated embryoid bodies did not express PDX-1.

Because expression of the PDX-1 gene is associated with the formation of beta islet cells, these results suggest that beta islet cells may be one of the cell types that spontaneously differentiate in the embryoid bodies.

The researchers now think that nerve growth factor may be one of the key signals for inducing the differentiation of beta islet cells and can be exploited to direct differentiation in the laboratory. Complementing these findings is work done by Jon Odorico of the University of Wisconsin in Madison using human embryonic cells of the same source. In preliminary findings, he has shown that human embryonic stem cells can differentiate and express the insulin gene.

More recently, Itskovitz-Eldor and his Technion colleagues further characterised insulin-producing cells in embryoid bodies. The researchers found that embryonic stem cells that were allowed to spontaneously form embryoid bodies contained a significant percentage of cells that express insulin.

Based on the binding of antibodies to the insulin protein, Itskovitz-Eldor estimates that 1 to 3 percent of the cells in embryoid bodies are insulin-producing beta-islet cells.

The researchers also found that cells in the embryoid bodies express glut-2 and islet-specific glucokinase, genes important for beta cell function and insulin secretion. Although the researchers did not measure a time-dependent response to glucose, they did find that cells cultured in the presence of glucose secrete insulin into the culture medium.

The researchers concluded that embryoid bodies contain a subset of cells that appear to function as beta cells and that the refining of culture conditions may soon yield a viable method for inducing the differentiation of beta cells and, possibly, pancreatic islets.

Taken together, these results indicate that the development of a human embryonic stem cell system that can be coaxed into differentiating into functioning insulin-producing islets may soon be possible.

Future Directions

Ultimately, type 1 diabetes may prove to be especially difficult to cure, because the cells are destroyed when the body's own immune system attacks and destroys them. This autoimmunity must be overcome if researchers hope to use transplanted cells to replace the damaged ones.

Many researchers believe that at least initially, immunosuppressive therapy similar to that used in the Edmonton protocol will be beneficial. A potential advantage of embryonic cells is that, in theory, they could be engineered to express the appropriate genes that would allow them to escape or reduce detection by the immune system.

Others have suggested that a technology should be developed to encapsulate or embed islet cells derived from islet stem or progenitor cells in a material that would allow small molecules such as insulin to pass through freely, but would not allow interactions between the islet cells and cells of the immune system. Such encapsulated cells could secrete insulin into the blood stream, but remain inaccessible to the immune system.

Before any cell-based therapy to treat diabetes makes it to the clinic, many safety issues must be addressed. A major consideration is whether any precursor or stem-like cells transplanted into the body might revert to a more pluripotent state and induce the formation of tumours. These risks would seemingly be lessened if fully differentiated cells are used in transplantation.

But before any kind of human islet-precursor cells can be used therapeutically, a renewable source of human stem cells must be developed. Although many progenitor cells have been identified in adult tissue, few of these cells can be cultured for multiple generations. Embryonic stem cells show the greatest promise for generating cell lines that will be free of contaminants and that can self renew. However, most researchers agree that until a therapeutically useful source of human islet cells is developed, all avenues of research should be exhaustively investigated, including both adult and embryonic sources of tissue.

Assessing Human Stem Cell Safety

The isolation of human stem cells offers the promise of a remarkable array of novel therapeutics. Biologic therapies derived from such cells—through tissue regeneration and repair as well as through the targeted delivery of genetic material—are expected to be effective in the treatment of a wide range of medical conditions. Efforts to analyse and assess the safety of using human stem cells in the clinical setting are vitally important to this endeavour. Transplanted human stem cells are dynamic biological entities that interact intimately with—and are influenced by—the physiology of the recipient. Before they are transplanted, cultured human stem cells are maintained under conditions that promote either the self-renewing expansion of undifferentiated progenitors or the acquisition of differentiated properties indicative of the phenotype the cells will assume.

After incompletely differentiated human stem cells are transplanted, additional fine-tuning occurs as a consequence of instructions received from the cells' physiologic microenvironments within the recipient. The capabilities to self-renew and differentiate that are inherent to human stem cells point simultaneously to their perceived therapeutic potential and to the challenge of assessing their safety.

Assessing human stem cell safety requires the implementation of a comprehensive strategy. Each step in the human stem cell development process—beginning with identifying and evaluating suitable human stem cell sources must be carefully scrutinised. Included in this global assessment are the derivation, expansion, manipulation, and characterisation of human stem cell lines, as well as preclinical efficacy and toxicity testing in appropriate animal models. Being able to trace back from the cell population prepared for transplantation to the source of the founder human stem cells also allows each safety checkpoint to be connected, one to the other.

Weaving a Stem Cell Safety Net

A diversity of opinion exists among researchers about the feasibility of initiating pilot clinical studies using human stem cells. Some are of the view that it is reasonable to expect within the next five years that human stem cells will be used in transplantation settings to replace dead or dying cells within organs such as the failing heart or that genetically modified human stem cells will be created for delivery of therapeutic genes. Others argue that a good deal more information about the basic biology of human stem cells needs to be accumulated before their therapeutic potential in humans can be assessed.

Clinical studies involving the transplantation of blood-restoring, or hematopoietic, stem cells have been under way for a number of years. Reconstituting the blood and immune systems through stem cell transplantation is an established practice for treating hematological malignancies such as leukemia and lymphoma. Transplanting hematopoietic stem cells resident in the bone marrow or isolated from cord blood or circulating peripheral blood is used to counter the destruction of certain bone marrow cells caused by high-intensity chemotherapeutic regimens used to battle various solid tumours. Moreover, clinical trials are being conducted to assess the safety and efficacy of using hematopoietic stem cell transplantation to treat various autoimmune conditions including multiple sclerosis, lupus, and rheumatoid arthritis. Although precedents exist for the clinical use of human stem cells, there is considerable reluctance to proceed with clinical trials involving human stem cells derived from embryonic and fetal sources. This hesitancy extends to adult human stem cells of non-hematopoietic origin, even though, by contrast, their plasticity is generally considered to be lower than that of their embryo- and fetus-derived counterparts. For human stem cells to advance to the stage of clinical investigation, a virtual safety net composed of a core set of safeguards is required.

Safety Assurance Begins with Adequate Donor Screening

Whether human stem cells are of embryonic, fetal, or adult origin, donor sources must be carefully screened. Routine testing should be done to guard against the inadvertent transmission of infectious diseases. Additionally, pedigree assessment and molecular genetic testing appear to be warranted. This is arguably the case when human stem cells intended for transplantation are derived from an allogeneic donor—that is, someone other that the recipient—and especially if the cells are obtained from a master cell bank that has been established using human embryonic stem or human embryonic germ cells.

The purpose of pedigree evaluation and/or genetic testing is to establish whether the human stem cells in question are suitable for use in the context of a particular clinical situation. For example, embryos derived from a donor with a family history of cardiovascular diseases may not be the best suited for the derivation of cardiac muscle cells intended to repair damaged heart tissue. Similarly, the use of molecular genetic analysis could detect a mutation in the gene for alpha-synuclein. This gene is known to be responsible for the rare occurrence of early onset Parkinson's Disease. Detecting such a genetic abnormality in neuronal progenitor cells derived from an established embryonic germ cell line could block the use of those cells as a treatment for a number of neurodegenerative conditions, including Parkinson's Disease.

The number of genes known to be directly responsible for causing disease or anomalous physiologic function is relatively small. Advances in techniques for identifying, isolating, and analysing genes, coupled with the wealth of information destined to become available as one outcome of the human genome sequencing projects, will raise this number. Considerably more will also be learned about how multiple gene products, each contributing an incremental quantity to the overall sum, predispose an individual to develop particular diseases. Clearly, it will eventually not be possible, or even necessary, to screen every source of human stem cells for the entire panoply of disease-associated genes. The screening of targeted genes will be conducted within the context of the relevant clinical population.

Using Controlled, Standardised Practices and Procedures for Establishing Cultured Human Stem Cell Lines Enhances Safety

To ensure the integrity, uniformity, and reliability of human stem cell preparations intended for clinical use, it is essential to demonstrate that rigorously controlled, standardised practices and procedures are being followed in establishing and maintaining human stem cell lines in culture.

Human stem cells from virtually every source other than blood-derived hematopoietic stem cells are maintained in tissue culture for some defined period of time. This is necessary to obtain a sufficient number of cells for use in clinical studies involving transplantation. Culturing human stem cells requires the use of formulated liquid media supplemented with growth factors and other chemical substances that promote cellular replication and govern the differentiation of the cultured human stem cells.

Since human stem cells are a dynamic, biological entity, failure to standardize procedures for maintaining and expanding cells in culture could result in unintended alterations in the intrinsic properties of the cells. The initial seeding density of the cells, the frequency with which the culture medium is replenished, and the density cells are permitted to achieve before subdividing will all affect the characteristics of human stem cells maintained in culture.

Altering the concentrations of supplemental growth factors and chemical substances, even switching from one supplier to another, may lead to changes in cell growth rate, expression of defining cell markers, and differentiation potential. Alterations in stem cell properties caused by the use of nonstandardised culture practices are likely to affect the behaviour and effectiveness of the cells once transplanted.

One particular concern is how safe it is to use serum derived from cows as a supplement to culture media. Due to the outbreak of bovine spongiform encephalopathy (BSE) in cattle herds, primarily those raised in the United Kingdom, only serum produced from cows reared in countries certified to be free of BSE should be used. Consumption of beef contaminated with the agent responsible for causing BSE has lead to the limited emergence of new variant Creutzfeldt-Jakob disease (nvCJD) in humans.

This disease results in the relentless destruction of brain tissue and is invariably fatal. Placing neural stem cells contaminated with the BSE infectious agent in a patient's nervous system to investigate cellular-replacement therapies for neurological disorders would be both irresponsible and devastating. Researchers are engaged in a vigorous effort to develop serum-free, chemically defined media that obviate risks associated with the use of bovine serum.

Alternatives to Culturing on a Feeder Layer of Animal Cells Improve Safety

An issue unique to the culturing of human embryonic stem and embryonic germ cells involves the use of mouse embryonic fibroblast feeder cells to keep the embryonic cells in a proliferating, undifferentiated condition. Human embryonic stem and embryonic germ cells are seeded directly onto a bed of irradiated mouse feeder cells. Transplanting into humans stem cell preparations derived from founder cells that have been in direct, intimate contact with nonhuman animal cells constitutes xenotransplantation—the use of organs, tissues, and cells derived from animals to treat human disease. The principal concern of xenotransplantation is the unintended transfer of animal viruses into humans.

Researchers are devoting considerable attention to developing culture conditions that do not use mouse feeder cells. In February of this year, scientists from Geron Corporation, a biotech company focusing on the development of embryonic stem cell technology for treating disease, presented findings at a scientific conference demonstrating that human embryonic stem cells can be maintained without mouse feeder cells.

Human embryonic stem cells seeded on a commercially available basement membrane matrix in media conditioned by feeder cells retain their proliferative potential and capacity to form all three embryonic germ layers (mesoderm, endoderm, and ectoderm). This suggests that human embryonic stem cells maintained in the absence of direct culture on a mouse feeder cell layer are comparable to human embryonic stem cells co-cultured with mouse feeder cells.

Detailed Characterisation of Human Stem Cell Populations Reinforces the Safety Net

Detailed characterisation of cell preparations intended for transplantation is critical to the development of human stem cells for clinical use. Identifying the cells that make up an human stem cell population intended for clinical study requires identifying cells exhibiting the desired phenotype within the preparation, as well as those that do not. This poses considerable challenges because human embryonic stem and embryonic germ cells have the capacity to give rise to all differentiated cell types, while adult human stem cells, though generally more restricted in their plasticity, are capable of generating all cell types that make up the tissue from which they were derived.

On the basis of the complex biological properties of human stem cells, including their potential to differentiate along multiple lineages and give rise to a variety of cell types, it is expected that the characterisation of stem cell preparations will require a panel of orthogonal assessments. Parameters that will prove useful in establishing identity include 1) cell morphology (visual microscopic inspection of cells to assess their appearance), 2) expression of unique cell-surface antigens (as is the case for $CD34^+$ hematopoietic stem cells), 3) characterisation of biochemical markers such as a tissue-specific enzymatic activity (e.g., enzymes that produce neurotransmitters for nerve cells), and 4) expression of genes that are unique to a particular cell type.

Further, analysis of the nuclear chromosomal karyotype may be used to assess genetic stability of established human embryonic stem and embryonic germ cell lines maintained in culture for extended periods of time. Continued development and standardisation of DNA microarray analysis (simultaneous screening for many genes) and proteomics (protein profiling) technologies will significantly enhance stem cell characterisation.

Rigorous and quantitative identification of cell types within a heterogeneous population of differentiating human stem cells provides the means to gauge purity of a cellular preparation. In turn, this permits evaluation of the extent to which purity of a human stem cell preparation predicts efficacy after transplantation. It is not necessarily the case that homogenous populations composed of a single cell type will be more effective as a cell-replacement therapy than mixed populations of cells.

It is conceivable that the reason differentiation of cultured stem cells obtained from the brain leads to formation of all the cell types found within the nervous system (namely, neurons, astrocytes, and oligodendrocytes) is that their coincidental presence is required to ensure maximum survival and functional capability. The interaction of various phenotypic cell types within a preparation of progenitor cells obtained after the controlled differentiation of cultured human embryonic stem cells is being actively investigated.

Once the purity profile has been established for a population of human stem cells generated using standardised procedures, deviations that occur outside what is expected due to normal biologic variation serve as a harbinger that significant, and possibly deleterious, changes may have occurred. Such alterations could reflect the introduction of genetic mutations as a consequence of culture conditions used to promote expansion and to induce differentiation of the progenitor cell population.

Before clinical studies involving human stem cell transplantation can be done, it is essential to demonstrate that human stem cell preparations possess relevant biological activity. The bioassay provides a quantitative measure of the potency of a cell preparation and ensures that cells destined for transplantation are not inert. Assays may be based on a biologic activity such as insulin release from pancreatic islet-like cells, glycogen storage by cells intended for regeneration of liver tissue, or synchronous contraction in the case of stem cell-derived cardiomyocytes to be used for repairing damaged heart muscle. When cells that have not acquired fully differentiated functionality are to be transplanted, it may be appropriate to use surrogate markers that predict the acquisition of the intended biologic activity upon further differentiation. (For example, counting tyrosine hydroxylase-expressing neural progenitor cells in a mixed population of cells intended to provide dopaminergic neurons for treating Parkinson's Disease could predict the acquisition of relevant biologic activity after transplantation.)

Assessment of Human Stem Cell Safety

A critical element of the safety net is the transplantation of human stem cells into animals to demonstrate that the therapy does what it is supposed to do ("proof of concept") and to assess toxicity. Admittedly, animal models of human disease are imperfect because most human maladies do not spontaneously occur in animals. Chemical, surgical, and immunologic methods are used to damage neurons; induce diabetes; simulate heart attacks, stroke, and hypertension; or compromise organ function. In situations when focal genetic lesions are known to cause disease, the creation of transgenic mouse colonies in which the culpable gene is either eliminated or over-expressed results in disease models that are capable of faithfully reproducing human-disease-specific pathologies.

Human stem cells must be transplanted into animal models of human disease. Transplantation of neural stem cells should demonstrate measurable evidence of efficacy in models of neurodegenerative disease, such as Parkinson's Disease, Huntington's disease, and amyotrophic lateral sclerosis (ALS), Alzheimer's disease, as well as spinal cord injury and stroke. Improved liver function after transplantation of hepatocyte precursors should be observed in an animal model of hepatic failure. Normalisation of blood insulin concentrations and amelioration of diabetic disease symptoms should result from the transplantation of pancreatic islet progenitors in a mouse model of diabetes. It is likely that in all cases, immunosuppression will be required due to immunologic incompatibility between humans and the animal model species (usually mouse or rat).

In addition to efficacy, evidence for anatomic and functional integration of transplanted human stem cells should be assessed. Human stem cells destined for transplantation may be tagged with a marker, such as green fluorescent protein, that allows transplanted cells to be readily identified upon histological examination. A similar approach should be used to evaluate the migration of transplanted human stem cells from the site of injection into adjacent and more distant tissues. The migration of transplanted human stem cells to a nontarget site and subsequent differentiation into a tissue type that is inappropriate for that anatomic location could be problematic. Questions about the use of embryonic compared with adult stem cells with respect to robustness and durability should be addressed in animal-transplantation models. Similarly, the issue of whether less differentiated cells will be more effective than more differentiated cells following transplantation should be investigated. Continued advancements in noninvasive imaging technologies, such as magnetic resonance imaging (MRI) and positron emission tomography (PET scanning), will allow these events to be observed in real time with reasonable resolution and without having to use large numbers of animals.

From the perspective of toxicology, the proliferative potential of undifferentiated human embryonic and embryonic germ cells evokes the greatest level of concern. A characteristic of human embryonic stem cells is their capacity to generate teratomas when transplanted into immunologically incompetent strains of mice. Undifferentiated embryonic stem cells are not considered as suitable for transplantation due to the risk of unregulated growth.

The question that remains is, at what point during differentiation does this risk become insignificant, if ever? Identifying the stage at which the risk for tumor formation is minimised will depend on whether the process of stem cell differentiation occurs only in a forward direction or is reversible. Before clinical trials are begun in humans, the issue of unregulated growth potential and its relationship to stem cell differentiation must be evaluated. It is essential that careful toxicology studies are performed that are of the appropriate duration and that involve transplantation into immunocompromised animals of undifferentiated or partially differentiated embryonic stem cells, as well as adult stem cells.

The Clinical Application of Pluripotent Cells: The Promise and the Challenges

Stem cells are distinguished from other cells by two characteristics: (1) they can divide to produce copies of themselves (self-renewal) under appropriate conditions and (2) they are pluripotent, or able to differentiate into any of the three germ layers: the endoderm (which forms the lungs, gastrointestinal tract, and interior lining of the stomach), mesoderm (which forms the bones, muscles, blood, and urogenital tract), and ectoderm (which forms the epidermal tissues and nervous system).

Pluripotent cells, which can differentiate into any mature cell type, are distinct from multipotent cells (such as hematopoietic, or blood-forming, cells) that can differ into a limited number of mature cell types. Because of their pluripotency and capacity for self-renewal, stem cells hold great potential to renew tissues that have been damaged by conditions such as type 1 diabetes, Parkinson's disease, heart attacks, and spinal cord injury. Although techniques to transplant multipotent or pluripotent cells are being developed for many specific applications, some procedures are sufficiently mature to be established options for care. For example, human hematopoietic cells from the umbilical cord and bone marrow are currently being used to treat patients with disorders that require replacement of cells made by the bone marrow, including Fanconi's anemia and chemotherapy-induced bone marrow failure after cancer treatment.

However, differentiation is influenced by numerous factors, and investigators are just beginning to understand the fundamental properties of human pluripotent cells. Researchers are gradually learning how to direct these cells to differentiate into specialised cell types and to use them for research, drug discovery, and transplantation therapy. However, before stem cell derivatives are suitable for clinical application, scientists require a more complete understanding of the molecular mechanisms that drive pluripotent cells into differentiated cells. Scientists will need to pilot experimental transplantation therapies in animal model systems to assess the safety and long-term stable functioning of transplanted cells. In particular, they must be certain that any transplanted cells do not continue to self-renew in an unregulated fashion after transplantation, which may result in a teratoma, or stem cell tumor. In addition, scientists must ascertain that cells transplanted into a patient are not recognised as foreign by the patient's immune system and rejected.

The state of the science currently lies in the development of fundamental knowledge of the properties of human pluripotent cells. The scientific capacity needs to be built, an understanding of the molecular mechanisms that drive cell specialisation needs to be advanced, the nature and regulation of interaction between host and transplanted cells needs to be explored and understood, cell division needs to be understood and regulated, and the long-term stability of the function in transplanted cells needs to be established.

Stem cells derived from an early-stage human blastocyst (an embryo fertilized in vitro and grown approximately five days in culture) have the capacity to renew indefinitely, and can theoretically provide an unlimited supply of cells. It is also possible to derive stem cells from non-embryonic tissues, including amniotic fluid, placenta, umbilical cord, brain, gut, bone marrow, and liver. These stem cells are sometimes called "adult" stem cells, and they are typically rare in the tissue of origin.

For example, blood-forming (hematopoietic) stem cell experts estimate that only 1 in 2000 to fewer than 1 in 10,000 cells found in the bone marrow is actually a stem cell. Because so-called "adult" stem cells include cells from the placenta and other early stages of development, they are more correctly termed "non-embryonic stem cells." Non-embryonic stem cells are more limited in their capacity to self renew in the laboratory, making it more difficult to generate a large number of stem cells for a specific experimental or therapeutic application.

Under normal conditions, non-embryonic stem cells serve as a repair pool for the body, so they typically differentiate only into the cell types found in the organ of origin. Moreover, there is little compelling evidence for trans-differentiation, whereby a stem cell from one organ differentiates into a mature cell type of a different organ. New discoveries may overcome these limitations of stem cells derived from non-embryonic sources, and research directed toward this goal is currently underway in a number of laboratories.

The Role of Cultured Cells in Understanding the Differentiation Process

Cultures of human pluripotent, self-renewing cells enable researchers to understand the molecular mechanisms that regulate differentiation, including epigenetic changes (traits that may be inherited that do not arise from changes in the DNA sequence) in the chromatin structure, developmental changes in gene expression, exposure to growth factors, and interactions between adjacent cells. Understanding these basic mechanisms may enable future scientists to mobilize and differentiate endogenous populations of pluripotent cells to replace a cell type ravaged by injury or disease.

Alternatively, scientists may some day be able to coax human pluripotent cells grown in the laboratory to become a specific type of specialised cell, which physicians could subsequently transplant into a patient to replace cells damaged by these same disease processes.

Scientists are gradually learning to direct the differentiation of pluripotent cell cultures into a specific type of cell, which can then be used as cellular models of human disease for drug discovery or toxicity studies. While it is not possible to predict the myriad ways that a basic understanding of stem cell differentiation may lead to new approaches for treating patients with cellular degenerative diseases, some avenues can be theorised. For example, in the case of Huntington's disease, a fatal neurodegenerative disorder, one could imagine that pluripotent cells derived from an embryo that carries Huntington's disease and differentiated into neurons in culture could be used to test drugs to delay or prevent degeneration.

Despite the incredible growth in knowledge that has occurred in stem cell research within the last couple of decades, investigators are just beginning to unravel the process of differentiation. Human pluripotent cell lines are an essential tool to understand this process and to facilitate the ultimate use of these cells in the clinic. To provide background on this fundamental topic, this article reviews the various potential sources and approaches that have been used to generate human pluripotent and multipotent cell lines, both of embryonic and non-embryonic origin.

Establishing Human Pluripotent Stem Cell Lines from Embryonic or Fetal Tissues

Currently, at least six embryonic sources have been used to establish human pluripotent stem cell lines. All approaches involve isolation of viable cells during an early phase of development, followed by growth of these cells in appropriate culture medium. The various sources of these initial cell populations are discussed in brief below. It should be noted that the manipulation and use of embryonic tissues has raised a number of ethical issues.

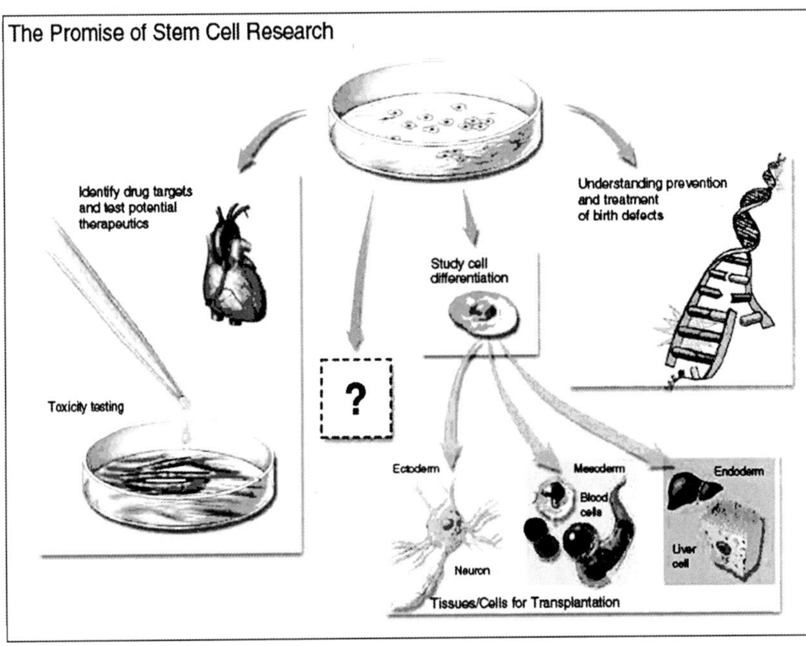

Figure: *The Promise of Stem Cell Research*

Stem cell research provides a useful tool for unravelling the molecular mechanisms that determine the differentiation fate of a pluripotent cell and for understanding the gene expression properties and epigenetic modifications essential to maintain the pluripotent state.

In the future, this knowledge may be used to generate cells for transplantation therapies, whereby a specific cell population compromised by disease is replaced with new, functional cells. Differentiated derivatives of human pluripotent cells may also prove to be useful as models for understanding the biology of disease and developing new drugs, particularly when there is no animal model for the disease being studied. The greatest promise of stem cell research may lie in an area not yet imagined.

Traditional Human Embryonic Stem Cell (hESC) Line Generation

Drawing upon twenty years of communal expertise with mouse ES cells, and on human inner cell mass culture conditions developed by Ariff Bongso and colleagues, James Thomson and colleagues at the University of Wisconsin generated the first hESC lines in 1998 using tissue from embryos fertilized in vitro. This method uses embryos generated for in vitro fertilization (IVF) that are no longer needed for reproductive purposes.

During IVF, medical professionals usually produce more embryos than a couple attempting to start a family may need. Spare embryos are typically stored in a freezer to support possible future attempts for additional children if desired. It is estimated that there are approximately 400,000 such spare embryos worldwide. If these embryos are never used by the couple, they either remain in storage or are discarded as medical waste. Alternatively, these embryos can potentially be used to generate a hESC line.

To generate a hESC line, scientists begin with a donated blastocyst-stage embryo, at approximately five days after IVF. The blastocyst consists of approximately 150–200 cells that form a hollow sphere of cells, the outer layer of which is called the trophectoderm. During normal development, the trophoblast becomes the placenta and umbilical cord. At one pole of this hollow sphere, 30–50 cells form a cluster that is called the inner cell mass (ICM), which would give rise to the developing fetus. ICM cells are pluripotent, possessing the capacity to become any of the several hundred specialised cell types found in a developed human, with the exception of the placenta and umbilical cord.

Scientists remove the ICM from the donated blastocyst and place these cells into a specialised culture medium. In approximately one in five attempts, a hESC line begins to grow. Stem cells grown in such a manner can then be directed to differentiate into various lineages, including neural precursor cells, cardiomyocytes, and hematopoietic (blood forming) precursor cells.

However, hESC lines are extremely difficult to grow in culture; the cells require highly specialised growth media that contain essential ingredients that are difficult to standardize. Yet the culture conditions are critical to maintain the cells' self-renewing and pluripotent properties. Culture requires the support of mouse or human cells, either directly as a "feeder" cell layer or indirectly as a source of conditioned medium in feeder-free culture systems. The feeder cells secrete important nutrients and otherwise support stem cell growth, but are treated so they cannot divide. Although the complete role of these feeder cells is not known, they promote stem cell growth, including detoxifying the culture medium and secreting proteins that participate in cell growth. hESC lines used to produce human cells for transplantation therapies may need to be propagated on a human feeder cell layer to reduce the risk of contamination by murine viruses or other proteins that may cause rejection. Thus, hESC lines often grow only under highly specific culture conditions, and the identification of ideal growth conditions presents a challenge regardless of the source of the hESCs.

Furthermore, human ES cell cultures must be expanded using an exacting protocol to avoid cell death and to control spontaneous differentiation. Since a limited number of laboratories in the United States are growing these cells, there is a shortage of people well-versed in the art and science of successful hESC culture. In the short term, challenges of working with these cells include developing robust culture conditions and protocols, understanding the molecular mechanisms that direct differentiation into specific cell types, and developing the infrastructure to advance this scientific opportunity.

Once these challenges have been met, scientists will need to conduct transplantation studies in animal models (rodent and non-human primates) to demonstrate safety, effectiveness, and long-term benefit before stem cell therapies may enter clinical trials.

HESC Lines from Human Primordial Germ Cells

A second method for generating human pluripotent stem cell lines was published in 1998 by John Gearhart and coworkers at The Johns Hopkins Medical School.

These researchers isolated specialised cells known as primordial germ cells (PGCs) from a 5–7 week old embryo and placed these cells into culture. PGCs are destined to become either oocytes or sperm cells, depending on the sex of the developing embryo. The resulting cell lines are called embryonic germ cell lines, and they share many properties with ES cells. As with ES cells, however, PGCs present challenges with sustained growth in culture. Spontaneous differentiation, which hinders the isolation of pure clonal lines, is a particular issue. Therefore, the clinical application of these cells requires a more complete understanding of their derivation and maintenance in vitro.

HESC Lines from Dead Embryos

Embryos that stop dividing after being fertilized in vitro are not preferentially selected for implantation in a woman undergoing fertility treatment. These embryos are typically either frozen for future use or discarded as medical waste. In 2006, scientists at the University of Newcastle, United

Kingdom, generated hESC lines from IVF embryos that had stopped dividing. These scientists used similar methods as described under "Traditional hESC Line Generation" except that their source material was so-called "dead" IVF embryos. The human stem cells created using this technique behaved like pluripotent stem cells, including producing proteins critical for "stemness" and being able to produce cells from all three germ layers. It has been proposed that an IVF embryo can be considered dead when it ceases to divide. If one accepts this definition, such an embryo that "dies" from natural causes presumably cannot develop into a human being, thereby providing a source to derive human ES cells without destroying a living embryo.

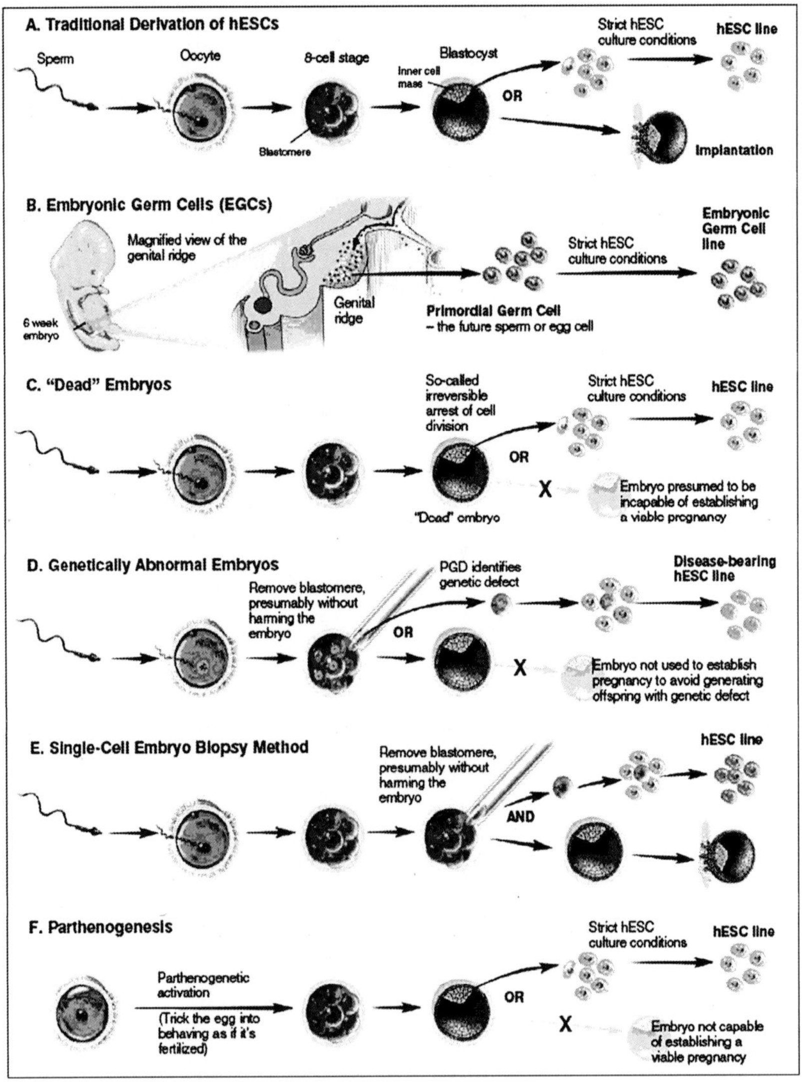

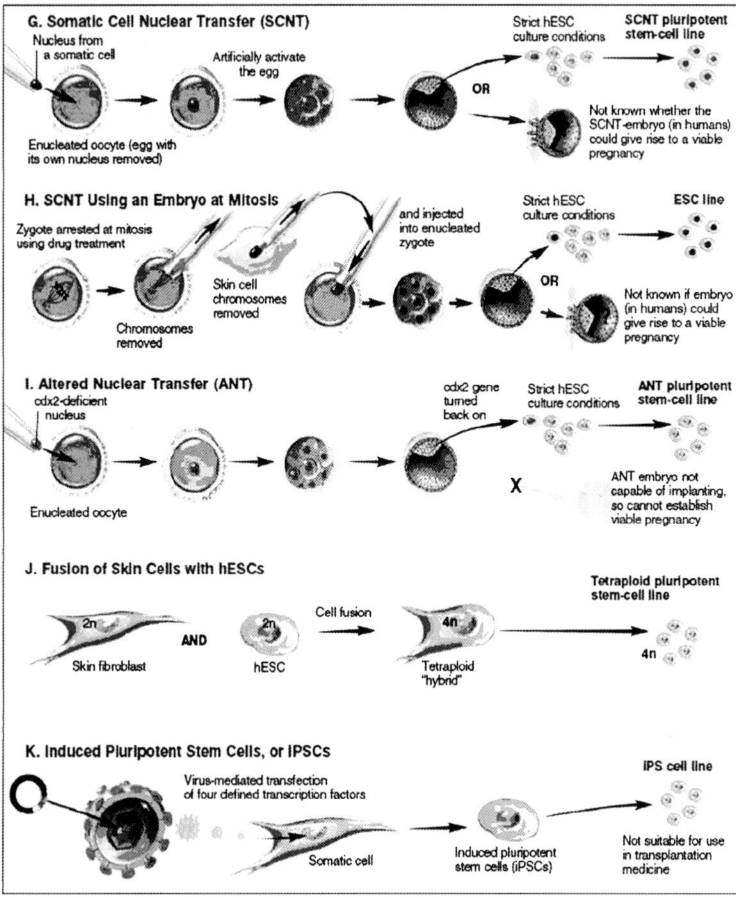

Figure: *Alternative Methods for Preparing Pluripotent Stem Cells.*

HESC Lines from Genetically Abnormal Embryos

Couples who have learned that they carry a genetic disorder sometimes use pre-implantation genetic diagnosis (PGD) and IVF to have a child that does not carry the disorder. PGD requires scientists to remove one cell from a very early IVF human embryo and test it for diseases known to be carried by the hopeful couple. Normally, embryos identified with genetic disorders are discarded as medical waste. However, Dr.Yuri Verlinsky and colleagues have capitalised on these embryos as a way to further our understanding of the diseases they carry by deriving hESC lines from them.

These stem cell lines can then be used to help scientists understand genetically-based disorders such as muscular dystrophy, Huntington's disease, thalessemia, Fanconi's anemia, Marfan syndrome, adrenoleukodystrophy, and neurofibromatosis.

HESC Lines from Single Cell Embryo Biopsy

In 2006, Dr. Robert Lanza and colleagues demonstrated that it is possible to remove a single cell from a pre-implantation mouse embryo and generate a mouse ES cell line. This work was based upon their experience with cleavage-stage mouse embryos. Later that same year, Dr. Lanza's laboratory reported that it had successfully established hESC lines from single cells taken from pre-implantation human embryos.

The human stem cells created using this technique behaved like pluripotent stem cells, including making proteins critical for "stemness" and producing cells from all three germ layers. Proponents of this technique suggest that since it requires only one embryonic cell, the remaining cells may yet be implanted in the womb and develop into a human being.

Therefore, scientists could potentially derive human embryonic stem cells without having to destroy an embryo. However, ethical considerations make it uncertain whether scientists will ever test if the cells remaining after removal of a single cell can develop into a human being, at least in embryos that are not at risk for carrying a genetic disorder. Moreover, it is unclear whether the single cell used to generate a pluripotent stem cell line has the capacity to become a human being.

HESC Lines Created Via Parthenogenesis

Parthenogenesis is the creation of an embryo without fertilizing the egg with a sperm, thus omitting the sperm's genetic contributions. To achieve this feat, scientists "trick" the egg into believing it is fertilized, so that it will begin to divide and form a blastocyst. In 2007, Dr. E.S. Revazova and colleagues reported that they successfully used parthenogenesis to derive hESCs.

These stem cell lines, derived and grown using a human feeder cell layer, retained the genetic information of the egg donor and demonstrated characteristics of pluripotency. This technique may lead to the ability to generate tissue-matched cells for transplantation to treat women who are willing to provide their own egg cells. It also offers an alternate method for deriving tissue-matched hESCs that does not require destruction of a fertilized embryo.

5

Stem Cell Research and Cloning

Stem cells are versatile cells in the body which are able both to reproduce themselves and to produce more specialised cells. As such, they are of great potential value in repairing and regenerating damaged cells and tissue. Many conditions are currently or potentially treatable with the use of stem cells, including Alzheimer's, Parkinson's, heart disease, stroke and diabetes.

Stem cells can be obtained from various sources. They can be obtained from the early human embryo, from the older human embryo or foetus, from the newborn baby (e.g. from the umbilical cord), from the older child and from the adult. As the individual develops, it is thought that stem cells become more committed to a particular destination in the body; however, some degree of flexibility appears to be retained. (We will be looking in the next section (2.1) at the clinical potential of adult versus embryonic cells.)

Embryonic Stem Cells

Embryonic stem cells can be obtained from the human embryo or foetus during or after an abortion, or after a miscarriage. They can also be obtained from the early human embryo after in vitro fertilization (IVF) or similar procedures. The IVF embryo may be 'spare', *i.e.* surplus to the infertile couple's requirements. Alternatively, the embryo may be specifically created for experimental use.

Cloning

Embryos may very soon be created for experimental use by means of cloning. In this procedure, the nucleus of an unfertilized ovum is replaced by the nucleus of a body cell from an existing human being. The ovum is then stimulated to create an embryo. As the nucleus contains almost all of the cell's genetic material, the embryo created is the clone or twin of the human being from whom the nucleus was taken, and could be used as a source of stem cells for research and eventual transplantation.

Cloning for research is sometimes described as therapeutic as opposed to reproductive cloning. In 'reproductive' cloning, the clone is transferred to the body of a woman and allowed to go to term. It should be noted that the cloning procedure itself is identical for 'reproductive' and 'therapeutic' cloning: the only difference lies in the purpose for which the clone would be created.

The terms 'reproductive' and 'therapeutic' in this context are misleading.1 'Therapeutic' cloning is not therapeutic for the clone, who will die in the course of its cells being taken when it is 5-7 days old. Such cloning *is*, moreover, reproductive, since it involves the creation of an embryo, although this embryo will not survive to the point of being born. Supporters of cloning have conceded that the clone embryo is a human organism: an early human life. In this Submission, we will be referring not to 'therapeutic' and 'reproductive' cloning, but to cloning for birth, and cloning for research/transplantation.

Scientific Issue

Adult Stem Cells

Until recently, it was thought that stem cells could rarely be found in adult tissue. However, this view has been conclusively disproved. The last twelve months have seen an explosion of information on adult stem cells (the term is often used to include stem cells from, for example, newborn babies). Adult stem cells have, in fact, been given to patients for decades in the course of bone marrow transplants, and new trials on patients using adult stem cells have had positive results.

It is clearly best if the cells used in transplantation can be taken from the patient him or herself, to avoid rejection by the body. While embryonic cells have been proposed as a means of avoiding rejection problems, even early embryonic cells have surface molecules which can cause an immune response. Supporters and opponents of embryonic stem cell research are in agreement that the ultimate goal should be to use the patient's own cells.

There are various ways in which adult cells can be used. They can be taken from the patient, or a donor, and used without being modified, as in the case of bone marrow transplants for patients with cancer. Alternatively, stem cells from the patient can be subjected to gene therapy before being re-introduced, as in the case of children who were treated recently for Severe Combined Immune Deficiency.

Stem cells can be induced to carry out a new role in the body, as when a patient with heart disease was given stem cells from muscle in his leg, which then formed a different kind of muscle in his heart. It is known that adult bone marrow cells are particularly versatile, and can produce bone, cartilage, tendon, muscle, fat, liver and neural cells. Neural stem cells can also form other cell types, such as blood and muscle cells.

Finally, there are ways of treating stem cells while they are still in the body of the subject. In animal studies, positive results have been obtained by adding the proper growth signal to the injured brains of rats, so that their own stem cells could proliferate.

Treatments of human patients using adult cells are already yielding results. Successful treatments have been carried out on children with cartilage defects, patients with corneal scarring, and patients with lupus, systemic sclerosis and rheumatoid arthritis. A number of cancers have been successfully treated using adult cells, including metastatic retinoblastoma, which has a poor prognosis with conventional treatments.

Embryonic Stem Cells

How do these achievements using adult cells compare with those made using embryonic cells? It should be noted that cells from early embryos have not so far been used on patients, and appear too unspecialised to control, unless modified in some way. We are some years away from any treatment using early embryonic cells.

Where stem cells from foetuses have been transplanted into patients along with other tissue the results have not been altogether positive. Even foetal cells can be difficult to control, and it is feared that embryonic stem cell transplants could give rise to cancer. In one case, a man with Parkinson's died after a transplant of foetal cells; it was later found that these cells had given rise to bone, skin and hair in the patient's brain.

A recent clinical trial did find some improvement in younger Parkinson's patients (those aged 60 or less). However, in 15% of those treated the patient developed permanent uncontrollable movements such as writhing, head-jerking and constant chewing. One patient now has to be tube-fed, so severe are his symptoms.

In animal experiments, there have been some favourable results using embryonic cells. However, there are technical problems involved in keeping human embryonic stem cells alive, and in making them differentiate along the right lines. A trial in which cells from human embryos were transplanted into rats found that these cells did not readily differentiate into brain cells. Instead, they stayed together in a disorganised cluster, and nearby cells began to die.

While it is too early to say whether embryonic or adult stem cells will ultimately prove more effective, there is more current evidence of usefulness in the case of adult cells. Statements to the contrary may owe more to external pressures - political or commercial - than to the scientific data.

To quote the journal *Science* in December 2000, 'the human embryonic stem cells and fetal germ cells that made headlines in November 1998 because they can, in theory, develop into any cell type have so far produced relatively modest results. Only a few papers and meeting reports have emerged from the handful of labs that work with human pluripotent cells.

Cloning for Birth

Cloning for birth has been carried out in various species of mammal, including sheep, cows, mice, pigs and goats. However, it is very difficult to produce a healthy animal by cloning. Problems arise at every point: in producing embryos by cell nuclear replacement, in bringing these embryos to term, and in obtaining healthy live-born offspring.

There is a high rate of miscarriage of clones, throughout the period of gestation, and a high rate of neonatal death. Major fluid retention in the gestating mother has also been observed.

Abnormalities in clones include developmental delays, heart and lung defects, malfunctioning immune systems, and excessive size. Genetic errors in the clone can cause unpredictable problems at any stage. Some mouse clones, after growing to the equivalent age in humans of thirty, suddenly become obese.

Despite these problems, there are those who wish to carry out cloning for birth in human beings. Fertility specialists Panos Zavos and Severino Antinori plan to clone for birth within the next two years. Zavos claims it will be possible to identify abnormalities at the embryo stage, so that abnormal embryos can be discarded. However, experts in animal cloning have suggested that this cannot be done, since there is no test for determining whether genes in the embryo have been properly reprogrammed.

Cloning for Transplantation

The high rate of serious abnormality in animal clones gives us reason to fear the effect on patients of transplanting into them genetically abnormal cells from an embryonic clone. In general, embryonic cells are unstable and difficult to control because of their pluripotency.

Cells from a cloned embryo, prone to genetic abnormality, raise still more difficulties. In any case, the great expense involved in cloning, and the need for a large number of human eggs, would be major impediments to its clinical use.

Status of the Human Embryo

What is a Human Moral Subject?

Morality and law are concerned with the respect that is due to human beings as moral subjects. People are not reducible to the sum of their physical parts or their biological drives. Mature and competent human beings can make decisions for themselves, for which they are held responsible both morally and legally. Furthermore, *all* human beings are members of the human community, whether or not they have come to full maturity, and all share a common humanity.

No human being is to be thought of as a *mere* physical object or a *mere* animal, for all possess a common dignity or significance that is not shared with any other animal. However, it is also important to stress that human beings *are* animals - of a particular sort. We should not think that the real person is a ghostly spirit inside the head (the mind or consciousness) or a demon who has taken possession of a body - as though the living human body were one thing and the person were something else.

This would degrade the human significance of the body and undermine the conviviality and bodily communication that help to constitute the human world. The Greeks defined a human being as a 'rational animal' and this seems to capture something of the essence of a human being, as long as we are careful not to define 'rational' in too narrow a fashion. Similarly the philosopher Boethius defined a subject (*persona*) as an individual being which has a rational nature: *persona est rationalis naturae individua substantia.* These definitions have the great virtue of simultaneously acknowledging the unity and the transcendence of the human individual. The human being is not two things joined together (a thinking thing and a physical thing) but one whole being.

It may seem that in the modern Western world, there is little danger of 'dualistic' ideas posing a real threat. However, there seems to be something in the Western way of life and in modern culture that encourages, on the one hand, a view of political and technological freedom as unlimited by any rules concerned with human nature and, on the other, a view of the body as purely mechanical and empty of any intrinsic human significance.

The human moral subject is the human being, who is, essentially, a very special animal. We should not be afraid to admit that we are animals, members of the species *homo sapiens*, and that our lives are the lives of *these* living human bodies. Human individuals are *biological* individuals. The question nevertheless arises: when and how does this human individual come to be? And does he or she possess full human moral status from the very beginning?

When does the Human Individual Begin?

Fertilization

In species which reproduce sexually, a new organism comes to be with the successful fusion of the gametes. This is a rule that applies not only to human beings but reflects the very nature of sexual reproduction. The offspring are genetically distinct from the parents, so that the moment of generation of the new individual is clearly defined. 'Fertilization is the process whereby the two sex cells (gametes) fuse together to create a new individual with genetic potential derived from both parents.'

If the distinction between gametes and embryo is clearly defined, so too is the continuity between the single-cell zygote and the multicellular adult organism. 'In nearly all cases the development of a multicellular organism begins with a single cell the fertilised egg, or zygote. While there are identifiable stages in the process of division, pattern formation and differentiation, the process itself is gradual and continuous. If one asks when and how development *begins*, the biological answer is clear: 'development begins with the fertilised egg, which is a single cell, giving rise to a number of smaller cells.

Progress within the science of developmental biology has done nothing to displace this most fundamental of starting points. The basic biological outlines of human development are not controversial: fertilization initiates a cascade of developmental events and is the universal *terminus a quo* for considering embryonic development. The fertilised ovum, termed a *zygote* (the single-cell phase) or a *conceptus* (the general term for the product of conception), is also termed an *embryo* from fertilization until the end of the second month when it is designated a *foetus*. Thus, from a biological perspective, the product of fertilization is a new organism, an embryo in the process of development. 'After these gametes combine, a new organism begins to develop.'

The Embryo and its Placenta

Biologists generally consider the new individual organism to begin with fertilization. Fertilization is a *process* that takes time - in human beings it takes about 24 hours from the first contact between the ovum and spermatozoon to the union of the chromosomes ('syngamy') and the first cell division - yet there is a point within this process when the most significant phase has been accomplished (when the process is essentially complete). From this point a new biological organism exists. However, some people have suggested that the embryo does *not* begin to exist when fertilization is complete. They point out that most of the cells of the *conceptus* (95% at the 64 cell stage) go to make the placenta and amniotic sac, and claim that, until the embryo 'proper' is clearly differentiated from these other structures, the human individual does not yet exist.

These critics are right in saying that the distinction between the placenta and the embryo 'proper' is not present immediately after conception. The placenta takes time to develop. However, they are mistaken in thinking that this distinction is a distinction between the embryo and something else. The placenta belongs to the embryo; it is an organ of the embryo by which it takes nourishment from its mother. If the amniotic sac and the placenta are understood as parts or organs of the embryo, then the appearance of these structures will be no more or less significant than the genesis of other embryonic structures. A new identifiable human individual already exists from the beginning, before these specialised structures start to appear.

Identical Twinning

A second phenomenon is more difficult to comprehend. Whereas most embryos develop into a single adult organism, a small number (0.3%) undergo division at an early stage so as to produce identical (or 'monozygotic') twins. The process of identical twinning is often thought to demonstrate conclusively that the early *conceptus* is not an individual. If a biological object is to be described as an individual organism it requires a certain level of integration, such that it is not just a collection of cells held together mechanically but a single integrated unit. Some claim that the phenomenon of identical twins shows that the early *conceptus* is not a unitary organism but only a loose collection of cells yet to develop into an organism or organisms.

A survey of textbooks of developmental biology shows that the embryo is more than an aggregate held together by mechanical forces. 'From the outset, the cells of the embryo are not only bound together mechanically, they are also coupled by gap junctions. Nor do biologists regard the phenomenon of twinning as something to be explained as the breaking up of a loose aggregate. They regard twinning as evidence of a *power* possessed by the embryo to redirect cells that normally would have contributed to only a portion of the embryo, a power they term regulation. This is a power of the embryo as a *whole*, rather than of individual *cells*. 'Since early vertebrate embryos show considerable capacity for regulation and many of the cells are not determined, this implies that cell-cell communication must determine cell fate.

Identical twinning is an exceptional event proving the rule that, in general (in 99.7% of cases), only one foetus develops from each zygote. Where identical twinning does occur, the twinning event (whatever this is) triggers the formation of one or more discrete organised wholes, each in the process of development. Each newly generated twin then continues as a well-formed individual. It does not form just half a foetus, nor does it grow wildly as a teratoma. Both in singletons and in identical twins, early human embryology shows a strong commitment to the development of distinct organised individuals.

Biologically speaking, the early embryo is a well-integrated whole which, even in the rare cases when it splits, does so in such a way that one or more well-integrated individuals are formed. However, it might be argued, do not even these rare exceptions undermine a principle that is supposed to be *absolute*? If the *conceptus* could split, or could be made to split, does this not show that it is not yet a genuine individual, even in those cases where it goes on to become an individual?

This argument relies on the principle that whatever is an individual *cannot* give rise to other individuals. This principle may once have seemed plausible; however, it is undermined by the very possibility of human cloning. In many living things, including some vertebrates, reproduction can occur asexually: by a parent-cell splitting to give two daughter-cells; or by a new organism budding off from a continuing parent. The suggestion that adult human beings might one day be cloned assumes that human beings *can* be reproduced asexually - without this compromising the individuality of the human being who generated the clone.

There is ample evidence for strong integration in the early embryo, both in the case of singletons and in the case of identical twins. From the perspective of biology it is clear that there is one individual which endures from the single cell stage until the death of the multicellular organism - except in the case of identical twins, who are natural clones produced early in development by asexual reproduction. Furthermore, as soon as these twins come to be, they *also* endure as individual organisms until their own deaths. The normal habit of biologists is to *count* embryos: to test them, select them and transfer them. In all these actions it is assumed to be unproblematic that even early embryos are a discrete number of individual living organisms such that they can readily be counted.

Do all Human Beings have Full Moral Status?

Human Rights

Human subjects are to be respected by others and protected by society. Though every concrete human society seeks to limit those who can be accounted full members or citizens, there is a certain minimum protection that is generally thought due to *every* human being, whether a full citizen or not. The classification of some categories of people - especially foreigners, criminals, slaves, the mentally ill, women, children, and the disabled - as less than human was explicit in many ancient and even in some modern societies. It was in reaction to the horrors of Nazism that the United Nations sought to frame a list of basic 'human rights' possessed by all people irrespective of their situation, simply on the basis that they are human beings. 'Human rights' language assumes the existence, or at least the possibility, of a universal community to which all human beings belong. Human rights are those things which are owed to a human being simply in virtue of being human.

The moral insight basic to the very idea of human rights is that the human moral community naturally extends to all human beings, so that the exclusion of some category of human beings as subhuman is arbitrary and unjust. Human beings are revealed to be moral subjects by their thoughts and their mature free decisions, but those who are too immature or incapacitated to exercise these capacities are not to be discriminated against on the basis of what they cannot do. Human dignity resides in what human beings *are* rather than what they can or cannot *do* at any particular moment. All forms of society favour the full-grown and the strong and the bright, but these favoured few should not consider themselves more worthy of protection than the weakest or least able.

Consider the protection due to a newborn infant. Not only the parents, but the great majority of people, recognize the child as more worthy of protection than the adult because more vulnerable. Physical abuse of small children rightly provokes a sense of horror or outrage, and few things are more shocking than the death of a child. The old and the sick are less immediately attractive than little children, but again, the sight of an old person who has been violently assaulted is viewed as monstrous precisely because the victim was so vulnerable.

The newborn infant, the mentally disabled adolescent and the Alzheimer's patient are no less human, and no less worthy of the protection of society, than the healthy worker or the brilliant student. They are all, equally, members of human society, which is, or ought to be, for the benefit of all. The human being who is mentally impaired or who has lost the use of his or her faculties is in a deprived state, but is not thereby subhuman.

The newborn infant cannot speak or understand any more than a dog or cat, but he or she is already human, and it is this humanity that is the basis for his or her developmental potential. The baby is human, someone's son or daughter, with a human future in store. The significance simply of being human is immediately recognised by men and women of good will and is the basis of efforts in civilized societies to protect children and to secure better treatment for the mentally impaired. Human infants *have* an interest in health, in a stable family and in a future, even before they can *take* an interest in these things.

Human Embryos

Many people who would strongly oppose discrimination on the basis of age or mental capacity, and who would defend the rights of children born with a disability, do not see the protection of five- or six-day-old human embryos in the same light. Whereas it is obvious to them that disabled children are human beings deserving protection, the same is not obvious in the case of human embryos. This may be for several reasons. Some do not consider human embryos to be human individuals. Others consider the allegedly high level of natural wastage of embryos points to their lack of worth. Finally there are many who simply cannot believe that the embryo is a human being because it appears to be just a ball of cells.

Essential to the proper understanding of the human embryo is that it is not just human life, but a human individual; that is to say, an integrated human organism. This has been argued at length in the previous section and, when contentious moral questions are not foremost in people's minds, there seems to be no problem in regarding human embryos as discrete biological individuals. The fact that one embryo occasionally produces two does not undermine the empirical observation that it was genuinely *one* before and genuinely gives rise to *two* after.

The number of ova which are successfully fertilized but without the embryo implanting is subject to much dispute with many conflicting estimates. It seems that there are abnormal fertilization giving rise to products which are so disordered as to lack all human developmental potential and which should not be regarded as embryos or as human individuals but as pseudo-embryos. On the other hand, it would be fair to say that, even among those conceptuses showing serious chromosomal abnormalities, many of these would be children suffering from severe disability, rather than pseudo-embryos.

Even if the normal level of wastage before implantation were 'only' 10-15% this would still amount to tens of thousands of embryos each year in Britain alone. Is this a cogent argument against attributing human status to the early embryo? High mortality may sometimes have the social effect of cheapening human life, but this is not a moral argument for taking this attitude oneself. In human beings, as in most animals, it is the young who are most vulnerable, and throughout history high rates of infant mortality have been the norm. This is still the case in many parts of the world. High rates of infant mortality probably contributed to the culture of the ancient world in which infanticide was regularly practised, but it did not justify the practice. Nature is sometimes cruel, but 'the survival of the fittest' is no guide to justice in society, and the prevalence of high mortality is no guide to the moral status of the human embryo.

It seems likely that neither the phenomenon of identical twins nor the rate of spontaneous embryo loss would have been given much weight if the individual in question had only *looked* more like a child. Newborn babies cannot exercise the human powers of speech and reason, but the potential they have is clearly rooted in their existing humanity. The mentally impaired, the sick and the old may be disabled and even disfigured, but they have human faces and human flesh and blood. The early embryo looks very different and does not easily evoke the concern or compassion evoked by a child or a comatose adult. Society recognises the human status of those with unrealised or frustrated human potential, but in every case this is based on an acknowledgment of the *visible* humanity of the person. It is precisely this which is in question in the case of the early embryo.

Appearance is important, and immediate human sympathy or compassion is an important weapon in overcoming cultural prejudice against certain categories of human being. However, it is the task of moral reasoning to criticize our immediate emotional reactions and expand our moral horizons. The early embryo, whether produced by fertilization, by natural cloning (identical twinning) or by artificial cloning (cell nuclear replacement), is a biological individual, a developing human being in the earliest stage of life who, if allowed to develop, could enjoy a human future. Killing a human embryo is unjust for the same reason that killing an unwanted newborn child is unjust: because an innocent human being is deliberately destroyed and robbed of a future.

Throughout history some human beings who were undeniably human, but who were different from others because of sex, race, age or disability, have been accorded a graded 'respect' without 'equality'. In each case the rhetoric of difference was invoked in defence of injustice. The basis of this injustice was the very existence of the category of a *human being who is not a member of human society*, a subhuman or *untermensch*, a category which was firmly rejected by the UN declaration of human rights. Human embryos are human individuals who, given time and opportunity, will grow into older human beings. They are no less human than any other human being but only younger and less aware, more vulnerable and less able to arouse our compassion.

Respecting the Embryo

Rights of the Embryo

We have seen that there are solid grounds for acknowledging the embryo as a human moral subject: an individual who has human interests and, therefore, basic human rights. What, then, are the basic rights of human beings as they apply to the embryo? We should distinguish, first of all, between the *positive* rights and the *negative* rights of human beings. Human beings have *positive* rights, that is, rights that other people make certain choices which concern them: to provide them with medical treatment, for example, or some other portion of the communal resources. While these rights are very important, it is clear that they do not apply in all situations where help might be needed.

It is easy to imagine situations where there are more patients than a doctor can treat, and no patient has more right to treatment than any other. It would therefore be wrong to say that each patient has an absolute right to be treated; the doctor must simply distribute resources as best he or she can. In contrast, the rights that human beings have *absolutely* are certain *negative* rights: rights that other people *not* deliberately harm them. For example, people have the right not to be intentionally killed (at least if they are innocent of violence or serious injustice). In the same way, people have the right not to have their bodies harmfully invaded; for example, not to have their organs harvested while they are alive.

As the embryo is a human moral subject, it too has the negative rights of an innocent human being. It has the right not to be deliberately killed, and not to have its body deliberately invaded in ways which do it lethal harm. The process of harvesting its cells, in the course of which the embryo dies, is no more permissible than extracting organs from a newborn child who dies as a result.

The embryo also has certain positive rights to shelter and nourishment from its mother. A pregnant mother has an extra reason to take care of her own health, for the sake of her unborn baby. She should also take reasonable steps to provide for the child's growth and development, by eating appropriately, for example.

Given that a mother has certain responsibilities to her child, it is clearly wrong for a mother (or father) to permit embryos to be conceived in vitro who will not have a chance of being born. Parents who have had children by IVF normally consider the embryo to be a child or potential child. It is clearly irresponsible to accept the status of parenthood in this way and at the same time neglect the 'spare' embryos who have been conceived.

Research on 'spare' IVF Embryos

In a typical course of IVF treatment, some embryos will be conceived who will be surplus to a couple's requirements. In the case of these embryos, who would otherwise be discarded, it may be thought that they are 'going to die anyway', and should therefore serve some purpose, rather than be wasted altogether. We should note that this approach would not be accepted in the case of other human beings who are threatened with imminent destruction. In the case of prisoners on death row, or patients who are dying but not yet dead, we can see how disrespectful and how dangerous it would be to use them as live organ 'donors'.

In any case, we need to ask why IVF embryos are created in such lavish numbers. Leaving aside other questionable aspects of the IVF process, it is surely unethical practice deliberately to produce greater numbers of embryos than will be transferred to the mother.

Potential parents seeking fertility treatment are in a vulnerable position and it is difficult for them to object to the use of their embryonic offspring for research into infertility. The dangers inherent in using patients for medical research into their own condition are well documented. The Helsinki declaration on research on human subjects, as emended at Edinburgh in 2000, states that 'special attention is required for those for whom the research is combined with care' (para. 8). The 1989 Hong Kong version was even clearer:

'The subjects should be volunteers - either healthy persons or patients for whom the experimental design is not related to the patient's illness.' Even if destructive embryo research were acceptable in principle, analogous considerations would mitigate against using any embryos supplied by patients undergoing a course of fertility treatment. The situation of need or dependence tends to undermine freedom of consent.

IVF for Research

One step beyond experimentation on surplus IVF embryos is the deliberate creation of IVF embryos for experimental use. Here there is no pretence that any embryo produced will be given a chance of survival. It is not a case of bringing some good out of the supposed accident of too many IVF embryos; it is creation for destruction.

Conceiving IVF embryos for research compounds the disrespect shown towards them. The industrial or commercial exploitation of embryos is a further violation of their rights. It involves treating embryos as livestock to be farmed, managed, traded and slaughtered. The degradation of the IVF embryo conceived for research lies not simply in its early destruction but in the whole context within which the embryo lives out its brief existence.

With the recent expansion of the legal grounds on which embryo research may be performed, IVF for research seems likely to become much more common in the future. As research expands it is inevitable that there will be calls to extend the period of embryonic development within which it is permitted, particularly if cells from early embryos prove difficult to manage. If research is permitted on human embryos, how long before it is seriously suggested that research should be extended to human foetuses?

Cloning for Research

Cloning for the purpose of research is an area of special concern. As with IVF for the purpose of research, it is intended not to allow the embryo to live longer than the experiment dictates. Already in the case of IVF even IVF to treat infertile couples the embryo risks being treated as a product or commodity, since the procedure itself involves a process of manufacture.

In the case of cloning, the method of production would be further dehumanised. The clone embryo would have no genetic parents, and would be created like a product to precise specifications. Having no parents in the normal sense, the clone would have no natural human protectors, and would be still more vulnerable than the IVF embryo today.

Cloning for research is the same procedure as cloning for birth, and already involves the decision to produce a human clone, although the clone will then be destroyed. The technical expertise developed in cloning for research will certainly assist those who want to clone for birth. Thus, if cloning for birth is regarded as objectionable, cloning for research must also be so regarded.

Conscience and Complicity

If the expansion of destructive research on IVF embryos and the initiation of cloning for research is allowed to go ahead, this will generate further moral problems concerning the issue of complicity in these activities. It is morally wrong not only to destroy human beings, but also to commission or authorize their destruction. Cloning and stem cell research create serious problems of conscience for doctors, patients, researchers and those asked to donate material to produce embryos for research. For example, a patient who supplies a cell for the purpose of creating a clone would be intending the destruction of the clone for the sake of harvesting its cells.

Complicity problems are not limited to cases where one *intends* the wrongdoing of others. Even those who do not intend an act of injustice can act wrongly themselves by giving the impression they condone it, if what they do is closely linked to such an act. Thus a patient might be acting wrongly if he or she accepted a stem cell treatment - even one which did not itself destroy embryos - if that treatment had been developed by means of the destruction of embryos.

If the *only* treatment developed for a serious medical condition is one that involves, or has involved, the creation and destruction of embryos, this will condemn conscientious physicians and patients to endure a cruel trial. Unless they act against their conscience - and do what they consider inhuman and barbaric - patients will suffer without hope of treatment and doctors will be unable to offer any alternative. This situation would be intolerable.

Consider alternative avenues of research where one might lead to successful treatment which would be universally welcomed, while the other, even if it led to successful treatment, would create problems of conscience for thousands of patients and physicians. If the research possibilities are as yet uncertain, then it is particularly unfair to pursue research that, if successful, will be unconscionable to a significant proportion of the population. This is especially so in an age of integrated national healthcare provision. It is invidious to alienate so many people from the healthcare system.

Foetal tissue can be obtained from foetuses who have miscarried spontaneously, with the permission of the parents. In the case of taking tissue from aborted foetuses, it is often claimed that the act of abortion can be separated from subsequent research on tissue from the foetus, in such a way as to make this research morally acceptable.

This claim is somewhat disingenuous: the mere presence of an intermediary between the abortionist and the researcher does not mean there is not a close link between their activities. In any event, with regard to stem cell research on early embryos, it is not possible to dissociate the embryo destruction from the research, since the embryo would be destroyed by the very scientist who does the experiment.

Abortion may be linked to stem cell research, not only as a way of obtaining cells directly, but also as a means of supplying ova for the creation of clone or IVF embryos. Adult women are currently reluctant to allow their ova to be used to create embryos for research. It has been suggested that ova from a young aborted female could be appropriated - thus treating her body and fertility as a mere means to an end. Not only would the embryo created from the foetal ovum be treated as if it were an object, but the embryo would be the offspring of a foetus who was herself treated in this way.

Cloning for Birth

Though cloning for birth does not involve the intention to destroy all the offspring created (and, in this respect, is better than cloning for research), it would nonetheless be seriously wrong for a host of other reasons.

The first objection to cloning for birth is the physical harm it would cause in cloned human beings. The experience of attempting to clone other mammals shows that the great majority of human clones would have genetically related disabilities. Some nuclear replacement would not result in a true human embryo, but in an entity with no developmental potential.

If embryos were created, most would have severe abnormalities, some of which would only become apparent late in pregnancy. Screening for genetic abnormality would 'eliminate' a proportion of such abnormalities, but only at the cost of eliminating the unborn child itself. Of the babies born alive, many would suffer from disabilities inflicted on them irresponsibly because of the means used to produce them. Nor is it only the child who would be abused in such experimental reproduction. Those women who chose to gestate a cloned child would risk the trauma of a very high rate of miscarriage, the trauma of being offered abortion when severe abnormalities were detected *in utero*, and the health risks of a difficult and atypical pregnancy. Whatever the motives of women who chose to take part, it would be irresponsible of the scientists to impose such risks on any human subject.

Even if cloning for birth was carried out in a large enough number of cases for the risks to be substantially reduced, this could only happen at the cost of many dead foetuses, many disabled infants and many pregnant women who had suffered excessively en route to making cloning safe.

Added to these serious but extrinsic reasons for objecting to cloning for birth is the intrinsic objection that it is demeaning to the cloned child. The reaction of unease or revulsion which is generated by the idea of creating multiple genetically identical copies of an existing person is not something which should be dismissed as irrational.

Our natural reactions can sometimes be a guide to important human values. In the case of human procreation there seems to be a symbolic as well as a practical value to the novelty generated by sexual reproduction. Family resemblance leaves symbolic room for novelty and individuality.

Human beings are not, of course, reducible to their genes, and a cloned child would be as distinct from his or her genetic original as identical twins are distinct from one another.

Nevertheless, deliberately producing people to resemble an existing person would make them *seem* like replacements, and this would undermine their quest for individuality and make them appear, to themselves and to society, as (replaceable) 'copies' rather than as (irreplaceable) originals. Identical twins are not deliberately produced to resemble each other, nor is there an age-gap between them. In contrast, a cloned child would always be a copy, deliberately produced so as to fulfill the potential already demonstrated by another.

The cloned child would be produced without genetic parents, as a copy of some existing person, perhaps someone who had recently died (a replacement child), or perhaps someone still alive (a 'designer baby'). Even if the clone survived to live a healthy life, he or she would live in the shadow of the person cloned, under pressure to be like the person he or she was made to resemble.

This compromises the interest of children in establishing a separate identity for themselves; it subjects the child's freedom unreasonably to the parents' desires. Irrespective of the psychological harm to the child which might or might not follow, cloning is intrinsically a form of excessive parental control, since the child is created as a copy of someone chosen by the parents.

Public Policy

General Principles

The implications of the embryo's status as a human moral subject do not stop with the duties of those individuals who are personally involved in the creation of an embryo. There is a wider question of how society should respond to issues surrounding research on human embryos. It is instructive to consider how society responds to the moral rights of children and the mentally impaired. The most basic of these rights, such as the right to life (that is, the right not to be deliberately attacked or lethally neglected) are seen as enforceable by law, while other rights are, at least, promoted as a matter of public policy.

Moral principles regulating medical or other research on human subjects were laid down at Helsinki in 1964 and subsequently revised at Tokyo (1975), Venice (1983), Hong Kong (1989), Somerset West, SA (1996) and Edinburgh (2000).

The Declaration of Geneva of the World Medical Association binds the physician with the words, "The health of my patient will be my first consideration," and the International Code of Medical Ethics declares that, "A physician shall act only in the patient's interest when providing medical care which might have the effect of weakening the physical and mental condition of the patient." In medical research on human subjects, considerations related to the well-being of the human subject should take precedence over the interests of science and society.

Medical research is subject to ethical standards that promote respect for all human beings and protect their health and rights. Some research populations are vulnerable and need special protection. The particular needs of the economically and medically disadvantaged must be recognised. Special attention is also required for those who cannot give or refuse consent for themselves, for those who may be subject to giving consent under duress, for those who will not benefit personally from the research and for those for whom the research is combined with care.

Several points may be noticed. First, in the 1964 declaration (and until 1989) these paragraphs, or their analogues, occurred in a section entitled 'Non-Therapeutic Biomedical Research Involving Human Subjects'. Research on a human subject without the subject's consent, where the research would cause serious harm to the subject, where the benefits to science and society take precedence over the life and well being of the subject, are forbidden on every count. If research on human embryos is correctly described as 'research on human subjects', then it stands in square contradiction to the principles laid down with so much care at Helsinki and elsewhere.

The right of vulnerable individuals not to be attacked always outweighs the long-term benefit to others of medical research. However, what are we to do in cases of doubt, where it is not clear whether the research is harmful, or where it is not even clear that there is a human being who will be harmed? Take the case in which it is not clear whether a patient is dead and where a research team wishes to use his organs. If the patient is dead then removing his organs will not harm him as there is no human individual present. However, if the patient is alive then removing his organs will kill him. In such cases the death of the patient must be established *beyond reasonable doubt* before it is legitimate to act.

In the case of human embryos, there is no social consensus as to their human or moral status. Many people, not all of whom have interests in the research community, do not believe that the early embryo is a human individual with an interest in being born, and in not being harmed. The matter is fiercely debated in respectable academic journals of philosophy, theology and social policy. The argument presented in this paper is a contribution to that debate, giving public reasons why the human embryo should be acknowledged as a human moral subject. However, even if there are other reasonable people who disagree on this matter, this does not nullify the existence of a large body of respectable and well-reasoned opinion holding that the human individual begins at the single cell stage. This body of academic opinion constitutes at the very least, a *reasonable doubt* within the community as a whole concerning the status of the embryo. If the case is unproven then the only ethical course is not to risk killing the innocent, even for the sake of benefit to others.

5.1.9 There is no reason why society should tolerate destructive research on human embryos, any more than similar research on older children. It is important not to set a precedent of lethal research on human subjects, however worthy the motive. Even if society as a whole is not sure if the embryo is a human individual, as a society we should at least admit that this is a strong possibility, and so exclude research on what may well be human beings.

Benefits and Alternatives

If there is a serious moral doubt, which continues to exist within the national and international community, then morally questionable research should not be permitted. However, Parliament and popular opinion has clearly been influenced by the argument that this research is *morally necessary* because it offers the only hope of finding a cure to serious diseases such as Parkinson's, Alzheimer's and multiple sclerosis. Yet the haste with which this decision was made gave little time to weigh the scientific evidence in what has been an unusually fast moving field. Far from supporting the necessity of embryonic stem cell research, the evidence from current research suggests that there are great difficulties with embryonic stem cell therapy, comparable in magnitude, if not greater, than those associated with adult stem cell therapy.

It is not the case that embryonic stem cell research offers a real chance in the near future of revolutionary therapy for the diseases mentioned. Nor is it the case that there are no alternative and promising avenues of research. The pluripotency of early embryonic cells, which is the reason they are thought attractive, is at the same time the greatest obstacle to their use. The flexibility of these cells makes them unpredictable in the kinds of cells they produce, in a way which could endanger the patient if they were ever used in transplants.

Clinical promise in this area is currently being shown by adult stem cell research. Use of adult stem cells poses no ethical problem in principle and, if the patient's own cells are used, avoids the problem of rejection. Such treatment, which is already underway, is morally acceptable to the vast majority of patients. To the extent that it is thought necessary, embryonic stem cell research can be carried on in morally acceptable ways, on cells taken from animal embryos or, with the permission of the parents, from foetuses who have spontaneously miscarried.

Researchers have a vested interest in stressing the prospects of their own research, but in the case of embryonic stem cell research there is reason to doubt these prospects and the necessity of adopting this particular approach. All things being equal, no scientist wishes to be restricted in which methods may be employed, but social responsibility sometimes makes legal limits right and important. We would suggest that the public has been much misled as to the prospects of embryonic stem cell research in relation to the (scarcely mentioned) prospects of adult stem cell research.

As there are ethical alternatives which offer similar or better immediate and/or long term prospects, the government should give serious reconsideration to the moral doubts which have been expressed and the public unease surrounding the use of embryos for research, and especially the licensing of human cloning for research purposes. Stem cell research on clone and other embryos has been made legal without necessity at the very time when adult stem cell research is making most progress.

Unacceptable Consequences

The statutory instrument which permitted cloning for research purposes was greeted with dismay by the rest of Europe. The prohibition of cloning recommended by the European Parliament covers cloning for research as well as cloning for birth. Indeed, if cloning for research is permitted it seems difficult to imagine how cloning for birth could be resisted.

It was only one week after the passing of the statutory instrument by the House of Lords that Severino Antinori hosted a conference in Rome to announce his intention to bring a clone baby to term. As so often, it seems that the technical challenge represented by an unprecedented step (in this case, cloning a human being) is incentive enough to impel any scientific project. The legitimacy of this form of research is a matter of global and not merely national concern, for research forbidden in one country will move to another.

It is imperative that the Select Committee inform itself on the grave misgivings expressed in other scientifically developed nations in Europe, America and Australasia, so as to reach a common mind on what are matters of global ethical importance. Not only would the acceptance of cloning for research make cloning for birth inevitable, in the judgment of many, but it would also make inevitable a further extension of research on human embryos beyond 14 days. There are places where logically and politically a line can be drawn, and cloning is one of few such lines. A ban on cloning is supported by a great mass of people internationally and could provide a bulwark against further erosion of ethical standards within medical research. If society is not able to resist the attraction of this research, when the benefits are uncertain and international opinion is against it, it is hard to imagine what, in the future, it would be able to resist.

If stem-cell research on human embryos and cloning for research go ahead, this will seriously compromise many physicians and patients if and when treatments based on this research become available. In contrast, if research funds are diverted to develop treatments that all can accept, patients and doctors will not be placed in this invidious situation.

Recommendation

It is not too late to reconsider this matter. Although the findings of the Select Committee have no weight in law, and the statutory instrument has already been passed, if the Committee were to argue strongly against permitting cloning for research, in particular, it would be difficult for the government to ignore its findings. The legislation might yet prove a dead letter, with the HFEA unwilling to give licenses, or further legislation might be introduced to prevent all human cloning.

A moratorium on using clone or IVF embryos in stem cell research would help adult stem cell research attract more of the available research funds so as further to prove its case as scientifically as well as ethically superior. Research on clone or other embryos contravenes basic principles of justice and seriously threatens international ethical standards in research. This is an opportunity to prevent a step being taken that need not and should not be taken. It is a question of having the prudence to know when a line must be drawn, and the courage to draw it.

Responses to Questions

The Select Committee has set out certain questions to which we here respond. However, we are not thereby endorsing these as the most pertinent questions which could be asked, and our answers cannot be taken as a summary of this Submission.

Do the Additional Purposes in the 2001 Regulations Raise Issues of Principle Different from the Purposes Specified in the 1990 Act?

Yes, destroying human embryos in stem-cell research leads to new questions of conscience (4.5.1-5; 5.3.4); furthermore, the interpretation of these regulations as allowing cloning for research raises the question of the ethical legitimacy of cloning (4.4.1-2; 4.6.1-6), and also the question of whether and how to promote alternative avenues of research (5.2.1-5). However, the most important issue of principle is one already compromised in the 1990 Act and which urgently needs to be reviewed; i.e. the legitimacy of producing and using human embryos for destructive research (4.1.1-4.3.3).

There is a range of different views worldwide on the acceptability of research on embryonic stem cells. What considerations underlie these differences? Do changes in the law here have implications for practice overseas and vice versa?

Those who accept embryonic stem-cell research either see no ethical objection to what seems to be a beneficial technical advance or else they have accepted the overriding necessity of this form of research.

Those who object do so for many different reasons. What they have in common is that they are unconvinced by claims of the imminence of the prospective benefits and/or by claims of the necessity of research using these means (2.1.1-2.2.5). Many also hold that there are ethical objections in principle to this kind of research (4.1.1-5.1.9; 5.3.1-4).

Have increased globalisation and other international commercial developments, in relation, for example, to e-commerce and patenting, changed the context of the debate in the UK? Would issues relating to research on embryos benefit from more attention at international level?

Yes scientific research and development is now an international endeavour and effective regulation and maintenance of ethical standards requires international co-operation. If one country allows unethical research then it will quickly attract those wishing to avoid restrictions placed on research in their own countries. Such international research tourism is already apparent with some fertility treatments, and cloning would surely follow this pattern. As international co-operation is essential, it is important to pay close attention to the concerns of international partners when framing domestic legislation in this area (5.3.1-2). What are the potential medical benefits of stem cell research? What is the most likely time-scale for realising them? What are the potential risks?

There are great potential benefits from stem-cell research for a range of diseases. Some patients are already benefiting from (adult) stem cell research; however, for other conditions it will take many years before we see results at the clinical level.

There are Differing Views on the Extent to Which Potential Treatments Could be Developed from Non-embryonic Stem Cells, Such as Adult and Umbilical cord Stem cells. What are the Advantages and Disadvantages of Working with These Alternative Sources of Stem Cells?

Treatments using adult and umbilical cord stem cells are already being carried out for some conditions (2.1.1-2.2.5). It is clearly best if the patient's own cells can be used to avoid rejection problems. Cloning has been mooted as a way to avoid these problems; however, the high rate of genetic abnormality found in animal clones casts doubt over the safety of cloning for transplantation.

Another advantage of adult stem cell research is that human ova (already a scarce resource in fertility treatment) are not required to generate adult stem cells. Adult stem cells are also more stable and less inclined to produce cancer or disorganised growths (2.2.1-4; 2.4.1).

What are the commercial interests involved in research in this area? Does increased commercial involvement create additional ethical difficulties? Biotechnology companies have invested heavily in cloning and embryonic stem cell patents. Commercial interests endanger the clarity of the ethical discussion as commercial benefits may distort the fair presentation of the medical benefits.

Human reproductive cloning (the transfer of an embryo created by cell nuclear replacement into a woman's uterus) is unlawful in the UK, and the Government has announced its intention of reinforcing this ban by specific primary legislation. Is there likely to be any pressure to resist such a ban? What are the principal ethical (and scientific) arguments against human reproductive cloning?

If it has proven politically difficult to block research on cloned embryos, even though this is unnecessary and unethical, it may also prove difficult to block requests that some clones be allowed to come to term to fulfill the desire of an infertile couple for a child.

Cloning for birth is objectionable: first, as it would involve the creation of hundreds of damaged embryos, most of whom would miscarry; second, as it would impose significant risks on the gestating mother; and third, as the clone would be produced asexually and according to a predetermined genetic plan. Cloning would deprive the child of natural parentage and genetic novelty and subject him or her unfairly to the desires of the commissioning parents. All cloning, including 'therapeutic cloning', is reproductive in the sense that it generates a new embryonic individual (1.3.2-4). In any primary legislation, human reproductive cloning should be defined as 'cloning for the purpose of implantation'. It is the initial generation of the clone embryo which should be prohibited, not the implanting of a clone embryo who already exists.

Does the extension of embryonic stem cell research, and, in particular, the technique of cell nuclear replacement therapy (therapeutic cloning) - designed to grow tissue for therapeutic purposes - increase the likelihood of human reproductive cloning in the future?

Yes it is clear to everyone that cloning for research facilitates cloning for birth both technically and politically (4.4.2).

It should, however, be made clear that generating clones who will then be destroyed in the course of 'harvesting' their cells is itself seriously wrong irrespective of the fact that it might lead to cloning for birth. Indeed, in its treatment of the clone embryo, 'therapeutic cloning' is a greater injustice than 'reproductive cloning' (3.3b.6-7).

Has the regulatory framework established by the 1990 Act operated effectively? Is it likely to remain adequate for the foreseeable future? Have any gaps appeared in the regime as a result of developments since 1990?

The 1990 Act presumed the legitimacy of creating embryos in the course of fertility treatment who would not be implanted, and even of creating embryos purely for the sake of destructive research. As these procedures are both unethical, the regulatory framework could never have 'operated effectively' in any ethical sense.

The intention of the statutory instrument seems to presuppose the existence of a significant gap, in that the 1990 Act forbade one technique of human cloning but did not envisage cloning by the 'Dolly' technique. It is remarkable that this gap should be thought a sufficient basis for the regulation of such an ethically contentious issue as the creation and use of cloned human embryos.

Do additional guidelines need to be developed to assist the Human Fertilization and Embryology Authority in issuing licences in accordance with the new Regulations? If so, what should the guidelines contain?

If fundamental ethical principles have already been compromised in the framing of the Regulations themselves, it is difficult to see what could be the ethical basis of any further guidelines. Additional specification could perform only a cosmetic or a rhetorical function, not the function of specifying in greater detail valid ethical principles.

How Scientists Clone Cells

On January 8, 2001, scientists at Advanced Cell Technology, Inc., announced the birth of the first -clone of an endangered animal, a baby bull gaur (a large wild ox from India and southeast Asia) named Noah. Although Noah died of an infection unrelated to the procedure, the experiment demonstrated that it is possible to save endangered species through cloning.

Cloning is the process of making a genetically identical organism through nonsexual means. It has been used for many years to produce plants (even growing a plant from a cutting is a type of cloning). Animal cloning has been the subject of scientific experiments for years, but garnered little attention until the birth of the first cloned mammal in 1997, a sheep named Dolly. Since Dolly, several scientists have cloned other animals, including cows and mice. The recent success in cloning animals has sparked fierce debates among scientists, politicians and the general public about the use and morality of cloning plants, animals and possibly humans.

Producing Clones: Plant Life

Nature- has been cloning organisms for billions of years. For example, when a strawberry plant sends out a runner (a form of modified stem), a new plant grows where the runner takes root. That new plant is a clone. Similar cloning occurs in grass, potatoes and onions. People have been cloning plants in one way or another for thousands of years. For example, when you take a leaf cutting from a plant and grow it into a new plant (vegetative propagation), you are cloning the original plant because the new plant has the same genetic makeup as the donor plant. Vegetative propagation works because the end of the cutting forms a mass of non-specialised cells called a callus. With luck, the callus will grow, divide and form various specialised cells (roots, stems), eventually forming a new plant.

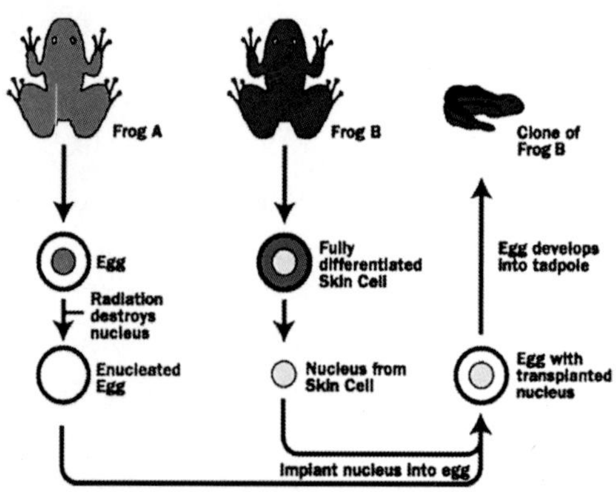

***Figure:** Gurdon's experiment to clone a frog*

More recently, scientists have been able to clone plants by taking pieces of specialised roots, breaking them up into root cells and growing the root cells in a nutrient-rich culture. In culture, the specialised cells become unspecialised (dedifferentiated) into calluses. The calluses can then be stimulated with the appropriate plant hormones to grow into new plants that are identical to the original plant from which the root pieces were taken.

This procedure, called tissue culture propagation, has been widely used by horticulturists to grow prized orchids and other rare flowers.

Producing Clones: Animal Kingdom

Plants are not the only organisms that can be cloned naturally. The unfertilized eggs of some animals (small invertebrates, worms, some species of fish, lizards and frogs) can develop into full-grown adults under certain environmental conditions usually a chemical stimulus of some kind. This process is called parthenogenesis, and the offspring are clones of the females that laid the eggs.

Another example of natural cloning is identical twins. Although they are genetically different from their parents, identical twins are naturally occurring clones of each other.

Scientists have experimented with animal cloning, but have never been able to stimulate a specialised (differentiated) cell to produce a new organism directly. Instead, they rely on transplanting the genetic information from a specialised cell into an unfertilized egg cell whose genetic information has been destroyed or physically removed.

In the 1970s, a scientist named John Gurdon successfully cloned tadpoles. He transplanted the nucleus from a specialised cell of one frog (B) into an unfertilized egg of another frog (A) in which the nucleus had been destroyed by ultraviolet light. The egg with the transplanted nucleus developed into a tadpole that was genetically identical to frog B.

While Gurdon's tadpoles did not survive to grow into adult frogs, his experiment showed that the process of specialisation in animal cells was reversible, and his technique of nuclear transfer paved the way for later cloning successes.

Dolly

In 1997, cloning was revolutionised when Ian Wilmut and his colleagues at the Roslin- Institute in Edinburgh, Scotland, successfully cloned a sheep named Dolly. Dolly was the first cloned mammal. Wilmut and his colleagues transplanted a nucleus from a mammary gland cell of a Finn Dorsett sheep into the enucleated egg of a Scottish blackface ewe. The nucleus-egg combination was stimulated with electricity to fuse the two and to stimulate cell division. The new cell divided and was placed in the uterus of a blackface ewe to develop. Dolly was born months later.

Dolly was shown to be genetically identical to the Finn Dorsett mammary cells and not to the blackface ewe, which clearly demonstrated that she was a successful clone (it took 276 attempts before the experiment was successful). Dolly has since grown and reproduced several offspring of her own through normal sexual means. Therefore, Dolly is a viable, healthy clone.

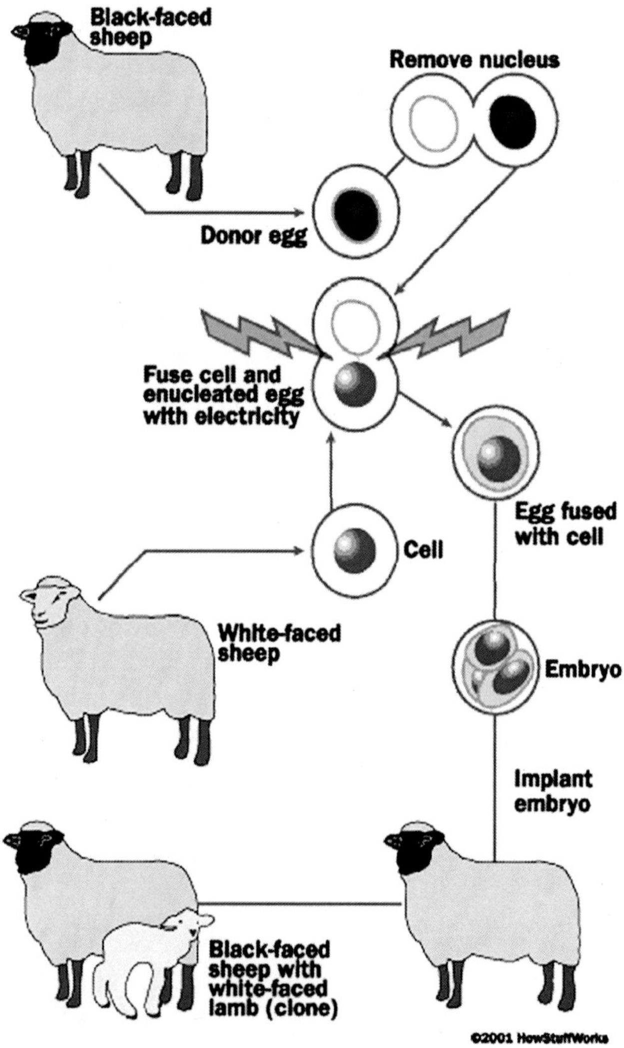

Figure: Diagram of the nuclear transfer procedure that produced the first cloned mammals

Since Dolly, several university laboratories and companies have used various modifications of the nuclear transfer technique to produce cloned mammals, including cows, pigs, monkeys, mice and Noah.

Why Clone?

The -main reason to clone plants or animals is to mass produce organisms with desired qualities, such as a prize-winning orchid or a genetically engineered animal for instance, sheep have been engineered to produce human insulin. If you had to rely on sexual reproduction (breeding) alone to mass produce these animals, then you would run the risk of breeding out the desired traits because sexual reproduction reshuffles the genetic deck of cards.

Other reasons for cloning might include replacing lost or deceased family pets and repopulating endangered or even extinct species. Whatever the reasons, the new cloning technologies have sparked many ethical debates among scientists, politicians and the general public. Several governments have considered or enacted legislation to slow down, limit or ban cloning experiments outright. It is clear that cloning will be a part of our lives in the future, but the course of this technology has yet to be determined.

The Cloning of Dolly

What makes Dolly different from identical twins is that she was grown from a cell taken from an ADULT animal! Many bright, well-respected scientists said it couldn't be done. Dr Ian Wilmut, who is in charge of the lab that created Dolly, admits that he had his doubts. However the hard work and imaginative thinking of his staff made it all possible.

How did they do it and what did they do?

First some background to teach you the basics of *developmental biology*.

An oocyte (pronounced "oh-oh-sight") is an unfertilized egg and it has no chance of developing into an animal unless it's fertilized. A recently fertilized egg is called a zygote (pronounced "zye-goat"). Funny how the last two letters in the alphabet describe the first stage of an individual animal. For example, a frog zygote normally divides and grows into a complete animal, a tadpole. Later that tadpole will develop into an adult frog.

A cell from a frog's gut should always remain a frog-gut-cell because it has "differentiated". Differentiation is the natural process whereby cells specialize into a certain kind of cell. As a frog embryo grows and develops its cells differentiate into nerve cells, blood cells, fat cells and many other different kinds of cells. That's what differentiation is all about. Differentiation is important because without it an animal could never be anything but a blob of unspecialised cells. As a mass of embryo cells divide and differentiate they "create" the animal.

This incredible process of differentiation turns zygotes into animals and it's all controlled by the genes. Although the exact process is still poorly understood, scientists agree that differentiation must have something to do with changes in the nucleus of cells. You may recall that the nucleus is the part of the cell containing the genetic material (the DNA all coiled up in organised structures called chromosomes).

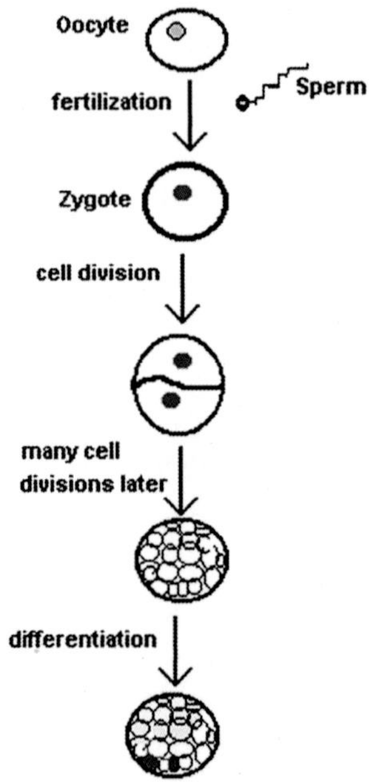

What do Frog Cells have to do with Dolly?

Like most scientific "breakthroughs" the earlier work done by others provided the foundation on which to try something new.

Decades ago a fellow named Gurdon developed the method of "nuclear transfer". This is a two step process.

First he used delicate needles and a good microscope to suck out the nucleus from a frog oocyte, producing an "enucleated oocyte". (That's an oocyte without a nucleus.) With the genetic material removed the enucleated oocyte would not divide or differentiate even when fertilized. That was no surprise. (A cell is nothing without its nucleus.)

But the results from Gurdon's second step shocked a lot of people! He used the same equipment and skill to transfer the nucleus from a frog's gut cell into an enucleated oocyte. That's nuclear transfer, the transfer of a nucleus from one cell to another, creating a "new cell" with a different nucleus. Many of these new cells which Gurdon created behaved like a zygote.

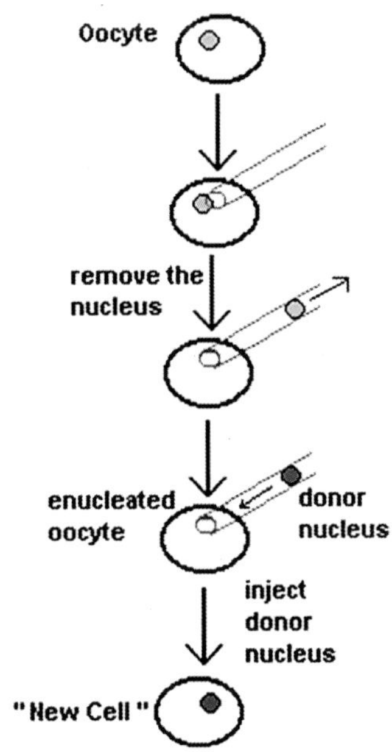

They divided and divided and divided just like a normal developing embryo, producing a ball of cells. And this ball of cells differentiated! Nerve cells, skin cells, blood cells appeared just as they would in a normal embryo. After the normal length of time Gurdon had tadpoles! Because the tadpoles had all come from the gut cells of the same adult, they all had the same genetic material. So they were all clones, identical twins of each other. But unlike normal identical twins they were made from differentiated cells. Gurdon had proven something that many scientists had argued about. He proved that differentiation was REVERSIBLE. Gurdon's method of nuclear transfer turned back the hands of time, in a developmental sense. Gurdon's method of nuclear transfer made clones from adult cells!

Naturally this got a lot of scientists thinking about cloning. But there were two problems.

First, Gurdon's nuclear transferred tadpoles never grew into frogs! Other folks repeated his experiments and got similar results. Nuclear transfer couldn't clone frogs from frog cells; all you got were tadpoles. No one knew why. Even today, no one knows why the tadpoles made by nuclear transfer die instead of growing into frogs. Weird.

The second problem was that Gurdon's method seemed to work only with frogs (or perhaps I should say "tadpoles"). When scientists tried nuclear transfer with mice, cattle or indeed any mammal, they got nowhere. The "new cells" sometimes divided a few times, but not for long and none of them differentiated properly. You just couldn't clone mammals. By the early 1980's most scientists accepted the idea that something very special allows frogs to be (partially) cloned (into tadpoles). Whatever that process was, it was not found in mammals. The textbooks made it very clear. Differentiation was (sort of) reversible in frogs but not in mammals.

So What exactly did the Scientists at the Roslin Institute do?

Well, Keith Campbell, a fellow working for Dr Wilmut, thought that maybe the cell cycle had something to do with this cloning trouble.

The cell cycle is often described as a circle of cell life and division, but I think that can be a bit confusing for some people, so let's try to remember that by "cycle" we mean it happens again and again.

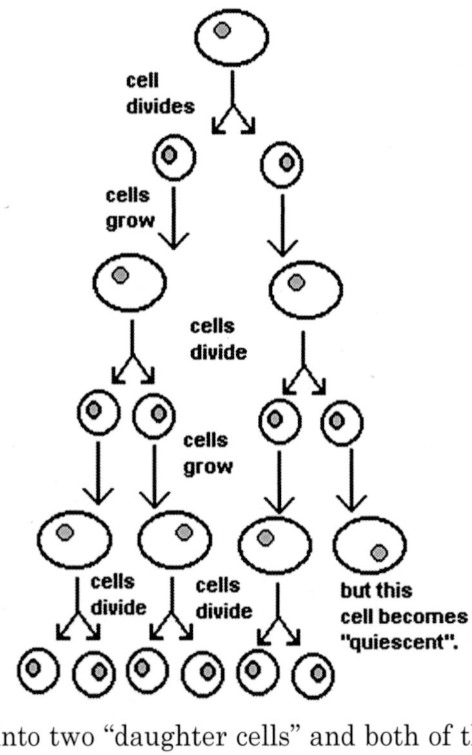

A cell divides into two "daughter cells" and both of these cells live, "eat", grow, copy their genetic material and divide again producing two more daughter cells. Because each daughter cell has a copy of the same genes in its nucleus, daughter cells are "clones" of each other just like identical twins. This "twining" goes on and on with each cell cycle. This is a natural process.

The cell cycle fascinates biologists. Very fast cell cycles occur during development causing a single cell to make many copies of itself as it grows and differentiates into an embryo. Some very fast cell cycles also occur in adult animals. Hair, skin and gut cells have very fast cell cycles to replace cells that naturally die. And cancer is a disease caused by cells cycling out of control. It's no wonder that biologists think the cell cycle is so important.

But there is a kind of "parking spot" in the cell cycle called "quiescence" (pronounced "kwee-S-ence"). A quiescent ("kwee-S-cent") cell has left the cell cycle, it has stopped dividing. Quiescent cells might reenter the cell cycle at some later time, or they might not. It depends on the type of cell. Most nerve cells stay quiescent forever. On the other hand, some quiescent cells may later re-enter the cell cycle in order to make more cells. (For example, when a young girl starts to develop breasts.)

Many biologists (including myself) thought that to make a clone you should transfer the nucleus from a fast dividing cell. It made sense because fast cycling cells are exactly what makes an embryo grow. Besides, the gut cells used to make the tadpole clones were fast cycling cells. Many biologists tried to make clones by transferring the nucleus from fast dividing cells but all of those experiments were unsuccessful. (I tried injecting the fast growing cells from chicken feathers into hen eggs in the hope of cloning birds, but it didn't work.)

Keith (Dr Campbell) thought about it in a different way. He wondered if a quiescent nucleus would be a better donor. True, it was not cycling (that's what makes it quiescent, by definition) but Keith thought maybe that's what the nucleus needs for it to be successfully transferred. Maybe the cell needs time to "rest" before starting to make a whole new animal. Maybe the nucleus needs time, lots of time, to get its DNA in order. Maybe...?

Maybe quiescent cells would work!

So they tried it with cells from sheep.

The folks at the Roslin Institute do a lot of work with sheep as part of their partnership with a company called Pharmaceutical Proteins Limited Therapeutics (PPL Therapeutics). Earlier they had made transgenic sheep (sheep with human genes transferred into them, but that's another story).

They used cells from an adult sheep's mammary (breast) glands for the "donor" nucleus. They grew the cells in tissue culture, an artificial situation that is commonly used in laboratories to grow large numbers of cells in bottles. Tissue culture allows scientists to fiddle with the cells and alter their characteristics. That is exactly what Dr Campbell did. He "starved" the cells of important nutrients and the cells stopped growing and dividing. They became quiescent. (Keith knew they would become quiescent when starved of nutrients because other researchers had proven that years ago; but few folks really cared because who needs quiescent cells?)

And then He Made Dolly?

Yes, but creating Dolly was not easy.

Using techniques similar to those used 20 years ago by Gurdon, Bill Ritchie (a technician working with Dr Campbell) removed the nucleus from an oocyte that was collected from a Scottish Blackface ewe.

(Ewes are female sheep. The Scottish Blackface breed is a common breed of sheep in Scotland easily identified by its black face.)

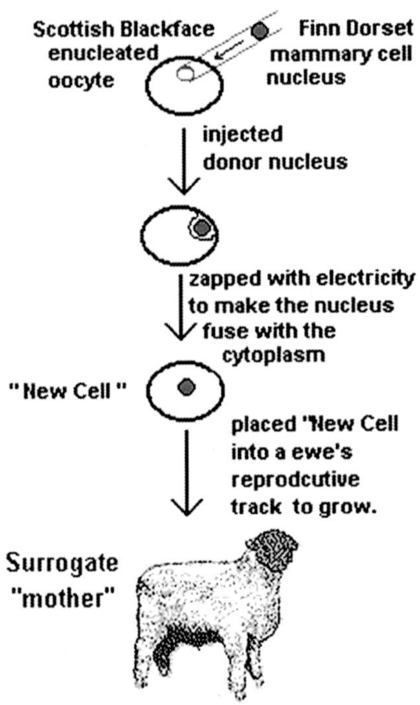

Oocytes have a "shell" of proteins and fibres (called the zona pellucida) and it is through this protective coat that Bill injected the nucleus from a quiescent mammary cell into the enucleated oocyte.

That cell nucleus was from a different breed of sheep called a Finn Dorset, which happens to be a pure white breed of sheep. He then used a tiny pulse of electricity to cause the new nucleus to fuse with the enucleated oocyte's cytoplasm. (Cytoplasm is the solution inside the cell.)

This electricity also helps "kick start" cells into "activity" so they are more likely to divide. This new, fused cell (containing the Finn Dorset mammary cell nucleus in the cytoplasm of a previously enucleated Blackface oocyte) was transferred into the reproductive "chamber" of a Blackface ewe (the same breed that provided the oocyte).

Bill and his fellow researchers than repeated this process 276 times! That's right, 276 times.

I told you this wasn't easy.

After 148 days, a normal length of time for the Finn Dorset breed of sheep, Dolly was born (5 July 1996).

As you can see she is a healthy, normal looking Finn Dorset. (Dolly's the wee one on the left) born to a Blackface ewe (her mom's on the right). This proves that Dolly wasn't the product of a sneaky mating; Dolly's Blackface mom could not produce a white faced sheep no matter who was the father. (It has to do with the genetics of sheep breeds.) But just to be sure, the scientists DNA "fingerprinted" Dolly and her "mom" and proved that Dolly's DNA matched the cells from the tissue culture, not the cells from the ewe that gave birth to her.

Dolly is a normal (Finn Dorset) sheep. Contrary to the reports in some of the trash newspapers, she has not eaten her keeper or her fellow sheep. She does not shoot laser beams out of her eyes or talk. Dolly is a friendly, normal, healthy sheep who enjoys being petted, especially if you have some food in your hand!

You may be surprised to learn that clones had been made at the Roslin Institute before, but those clones were made from the nucleus of embryo cells not adult cells.

6

Chemistry of Life

Basic Chemistry of Nucleic Acid

Living organisms are complex systems. Hundreds of thousands of proteins exist inside each one of us to help carry out our daily functions. These proteins are produced locally, assembled piece-by-piece to exact specifications. An enormous amount of information is required to manage this complex system correctly. This information, detailing the specific structure of the proteins inside of our bodies, is stored in a set of molecules called nucleic acids.

Adenosine Derivatives

The most common adenosine derivative is the cyclic form, 3'-5'-cyclic adenosine monophosphate, cAMP. This compound is a very powerful second messenger involved in passing signal transduction events from the cell surface to internal proteins, e.g. cAMP-dependent protein kinase, PKA. PKA phosphorylates a number of proteins, thereby, affecting their activity either positively or negatively.

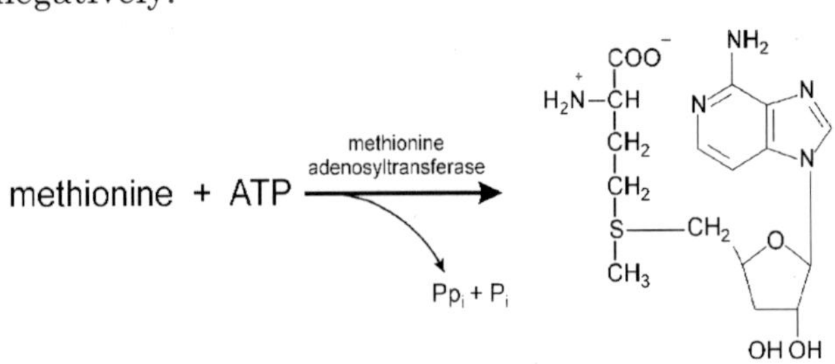

Figure: *Structure and Synthesis of S-adenosylmethionine.*

Cyclic-AMP is also involved in the regulation of ion channels by direct interaction with the channel proteins, e.g. in the activation of odorant receptors by odorant molecules. Formation of cAMP occurs in response to activation of receptor coupled adenylate cyclase. These receptors can be of any type, e.g. hormone receptors or odorant receptors.

S-adenosylmethionine is a form of activated methionine which serves as a methyl donor in methylation reactions and as a source of propylamine in the synthesis of polyamines.

Guanosine Derivatives

A cyclic form of GMP (cGMP) also is found in cells involved as a second messenger molecule. In many cases its' role is to antagonize the effects of cAMP. Formation of cGMP occurs in response to receptor mediated signals similar to those for activation of adenylate cyclase. However, in this case it is guanylate cyclase that is coupled to the receptor.

The most important cGMP coupled signal transduction cascade is that photoreception. However, in this case activation of rhodopsin (in the rods) or other opsins (in the cones) by the absorption of a photon of light (through 11-*cis*-retinal covalently associated with rhodopsin and opsins) activates transducin which in turn activates a cGMP specific phosphodiesterase that hydrolyzes cGMP to GMP. This lowers the effective concentration of cGMP bound to gated ion channels resulting in their closure and a concomitant hyperpolarization of the cell.

Nucleotide Derivatives in tRNAs

As many as 25% of the nucleotides found in tRNAs are modified post-transcriptionally. At least 80 different modified nucleotides have been identified at >60 different positions in various tRNAs.

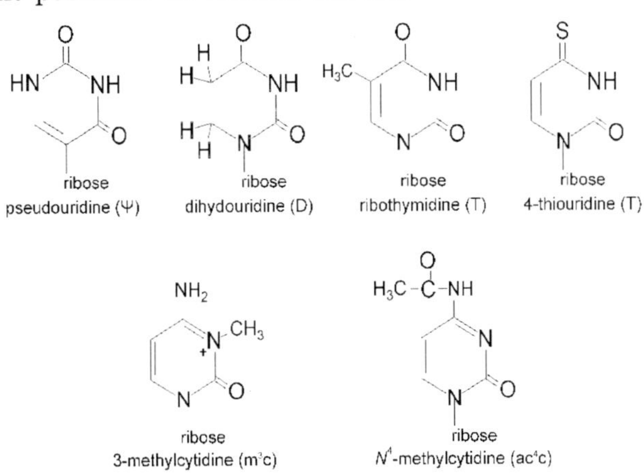

Figure: *Pyrimidine Derivatives*

The two most commonly occurring modified nucleotides are dihydrouridine (abbreviated D) and pseudouridine (abbreviated Ψ) each of which is found in a characteristic loop structure in tRNAs. Dihydrouridine is found in the D-loop (hence the name of the loop) and pseudouridine is found in the TΨC loop (hence the name of the loop.

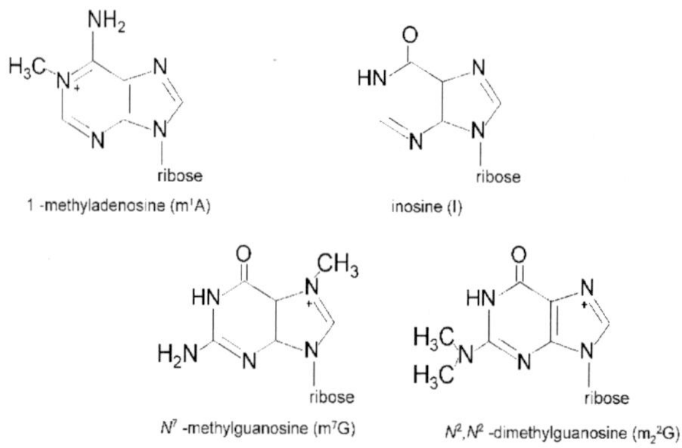

Figure: *Purine Derivatives*

Synthetic Nucleotide Analogs

Many nucleotide analogues are chemically synthesized and used for their therapeutic potential. The nucleotide analogues can be utilized to inhibit specific enzymatic activities. A large family of analogues are used as anti-tumour agents, for instance, because they interfere with the synthesis of DNA and thereby preferentially kill rapidly dividing cells such as tumour cells. Some of the nucleotide analogues commonly used in chemotherapy are 6-mercaptopurine, 5-fluorouracil, 5-iodo-2'-deoxyuridine and 6-thioguanine. Each of these compounds disrupts the normal replication process by interfering with the formation of correct Watson-Crick base-pairing.

Nucleotide analogs also have been targeted for use as antiviral agents. Several analogs are used to interfere with the replication of HIV, such as AZT (azidothymidine) and ddI (dideoxyinosine).

Several purine analogs are used to treat gout. The most common is allopurinol, which resembles hypoxanthine. Allopurinol inhibits the activity of xanthine oxidase, an enzyme involved in *de novo* purine biosynthesis. Additionally, several nucleotide analogues are used after organ transplantation in order to suppress the immune system and reduce the likelihood of transplant rejection by the host.

Polynucleotides

Polynucleotides are formed by the condensation of two or more nucleotides. The condensation most commonly occurs between the alcohol of a 5'-phosphate of one nucleotide and the 3'-hydroxyl of a second, with the elimination of H_2O, forming a phosphodiester bond. The formation of phosphodiester bonds in DNA and RNA exhibits directionality. The primary structure of DNA and RNA (the linear arrangement of the nucleotides) proceeds in the 5'—>3' direction. The common representation of the primary structure of DNA or RNA molecules is to write the nucleotide sequences from left to right synonymous with the 5'—>3' direction as shown:

5'-pGpApTpC-3'

Structure of DNA

Utilizing X-ray diffraction data, obtained from crystals of DNA, James Watson and Francis Crick proposed a model for the structure of DNA. This model (subsequently verified by additional data) predicted that DNA would exist as a helix of two complementary antiparallel strands, wound around each other in a rightward direction and stabilized by H-bonding between bases in adjacent strands.

In the Watson-Crick model, the bases are in the interior of the helix aligned at a nearly 90 degree angle relative to the axis of the helix. Purine bases form hydrogen bonds with pyrimidines, in the crucial phenomenon of base pairing. Experimental determination has shown that, in any given molecule of DNA, the concentration of adenine (A) is equal to thymine (T) and the concentration of cytidine (C) is equal to guanine (G). This means that A will only base-pair with T, and C with G. According to this pattern, known as Watson-Crick base-pairing, the base-pairs composed of G and C contain three H-bonds, whereas those of A and T contain two H-bonds. This makes G-C base-pairs more stable than A-T base-pairs.

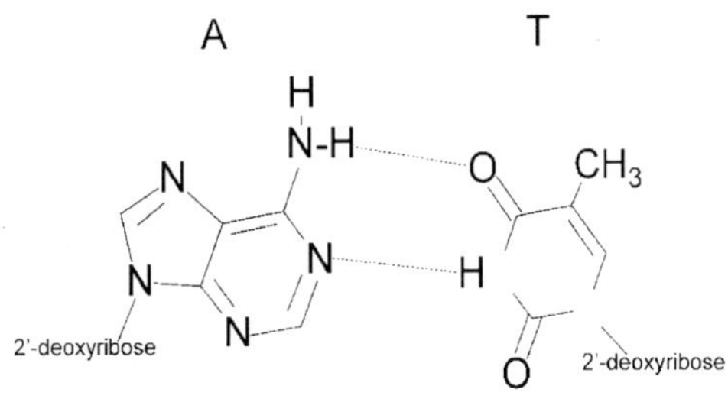

Figure: *A-T Base Pair*

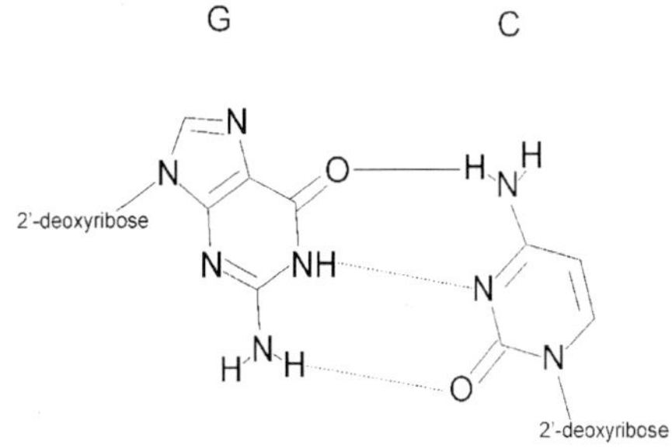

Figure: *G-C Base Pair*

The antiparallel nature of the helix stems from the orientation of the individual strands. From any fixed position in the helix, one strand is oriented in the 5'—>3' direction and the other in the 3'—>5' direction. On its exterior surface, the double helix of DNA contains two deep grooves between the ribose-phosphate chains. These two grooves are of unequal size and termed the major and minor grooves. The difference in their size is due to the asymmetry of the deoxyribose rings and the structurally distinct nature of the upper surface of a base-pair relative to the bottom surface.

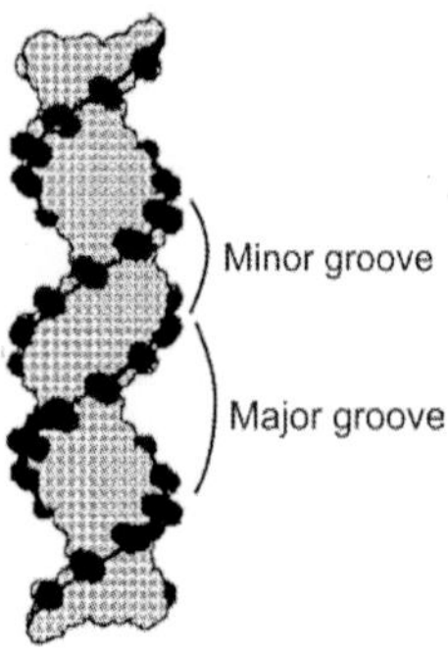

Figure: *Structure of B-DNA*

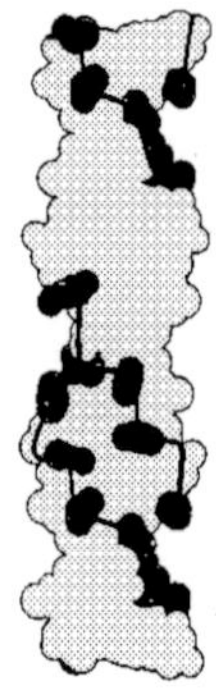

Figure: *Structure of Z-DNA*

The double helix of DNA has been shown to exist in several different forms, depending upon sequence content and ionic conditions of crystal preparation. The B-form of DNA prevails under physiological conditions of low ionic strength and a high degree of hydration. Regions of the helix that are rich in pCpG dinucleotides can exist in a novel left-handed helical conformation termed Z-DNA. This conformation results from a 180 degree change in the orientation of the bases relative to that of the more common A- and B-DNA.

Table: *Parameters of Major DNA Helices*

Parameters	A Form	B Form	Z-Form
Direction of helical rotation	Right	Right	Left
Residues per turn of helix	11	10	12 base pairs
Rotation of helix per residue (in degrees)	33	36	-30
Base tilt relative to helix axis (in degrees)	20	6	7
Major groove	narrow and deep	wide and deep	Flat
Minor groove	wide and shallow	narrow and deep	narrow and deep
Orientation of *N*-glycosidic bond	anti	anti	anti for Pyrimidines, syn for Purines
Comments		most prevalent within cells	occurs in stretches of alternating purine-pyrimidine base pairs

Thermal Properties of DNA

As cells divide it is a necessity that the DNA be copied (replicated), in such a way that each daughter cell acquires the same amount of genetic material. In order for this process to proceed the two strands of the helix must first be separated, in a process termed denaturation. This process can also be carried out in vitro. If a solution of DNA is subjected to high temperature, the H-bonds between bases become unstable and the strands of the helix separate in a process of thermal denaturation.

The base composition of DNA varies widely from molecule to molecule and even within different regions of the same molecule. Regions of the duplex that have predominantly A-T base-pairs will be less thermally stable than those rich in G-C base-pairs. In the process of thermal denaturation, a point is reached at which 50% of the DNA molecule exists as single strands. This point is the melting temperature (T_m), and is characteristic of the base composition of that DNA molecule. The T_m depends upon several factors in addition to the base composition. These include the chemical nature of the solvent and the identities and concentrations of ions in the solution.

When thermally melted DNA is cooled, the complementary strands will again re-form the correct base pairs, in a process is termed annealing or hybridization. The rate of annealing is dependent upon the nucleotide sequence of the two strands of DNA.

Analysis of DNA Structure

Chromatography

Several of the chromatographic techniques available for the characterization of proteins can also be applied to the characterization of DNA. The most commonly used technique is HPLC (high performance liquid chromatography). Affinity chromatographic techniques also can be employed. One common affinity matrix is hydroxyapatite (a form of calcium phosphate), which binds double-stranded DNA with a higher affinity than single-stranded DNA.

Electrophoresis

This procedure can serve the same function with regard to DNA molecules as it does for the analysis of proteins. However, since DNA molecules have much higher molecular weights than proteins, the molecular sieve used in electrophoresis of DNA must be different as well. The material of choice is agarose, a carbohydrate polymer purified from a salt water algae. It is a copolymer of mannose and galactose that when melted and re-cooled forms a gel with pores sizes dependent upon the concentration of agarose. The phosphate backbone of DNA is highly negatively charged, therefore DNA will migrate in an electric field. The size of DNA fragments can then be determined by comparing their migration in the gel to known size standards. Extremely large molecules of DNA (in the range of 30kb–10Mb) are effectively separated in agarose gels using pulsed-field gel electrophoresis (PFGE). This technique employs two or more electrodes, placed orthogonally with respect to the gel, that receive short alternating pulses of current. PFGE allows whole chromosomes and large portions of chromosomes to be analysed.

Carbohydrates

In many ways, our bodies can be thought of as chemical processing plants. Chemicals are taken in, processed through various types of reactions, and then distributed throughout the body to be used immediately or stored for later use. The chemicals used by the body can be divided into two broad categories: macronutrients, those substances that we need to eat regularly in fairly large quantities, and micronutrients, those substances that we need only in small amounts. Three major classes of macronutrients are essential to living organisms: carbohydrates, fats, and proteins

A carbohydrate is an organic compound which has the empirical formula $C_m(H_2O)_n$; that is, consists only of carbon, hydrogen and oxygen, with a hydrogen:oxygen atom ratio of 2:1 (as in water). Carbohydrates can be viewed as hydrates of carbon, hence their name. Structurally however, it is more accurate to view them as polyhydroxy aldehydes and ketones. The term is most common in biochemistry, where it is a synonym of saccharide. The carbohydrates (saccharides) are divided into four chemical groupings: monosaccharides, disaccharides, oligosaccharides, and polysaccharides.

In general, the monosaccharides and disaccharides, which are smaller (lower molecular weight) carbohydrates, are commonly referred to as sugars. While the scientific nomenclature of carbohydrates is complex, the names of the monosaccharides and disaccharides very often end in the suffix-ose. For example, blood sugar is the monosaccharide glucose, table sugar is the disaccharide sucrose, and milk sugar is the disaccharide lactose.

Carbohydrates perform numerous roles in living things. Polysaccharides serve for the storage of energy (e.g., starch and glycogen) and as structural components (e.g., cellulose in plants and chitin in arthropods). The 5-carbon monosaccharide ribose is an important component of coenzymes (e.g., ATP, FAD, and NAD) and the backbone of the genetic molecule known as RNA. The related deoxyribose is a component of DNA. Saccharides and their derivatives include many other important biomolecules that play key roles in the immune system, fertilization, preventing pathogenesis, blood clotting, and development. In food science and in many informal contexts, the term carbohydrate often means any food that is particularly rich in the complex carbohydrate starch (such as cereals, bread and pasta) or simple carbohydrates, such as sugar (found in candy, jams and desserts).

Structure

Formerly the name "carbohydrate" was used in chemistry for any compound with the formula $C_m(H_2O)_n$. Following this definition, some chemists considered formaldehyde CH_2O to be the simplest carbohydrate, while others claimed that title for glycolaldehyde Today the term is generally understood in the biochemistry sense, which excludes compounds with only one or two carbons.

Natural saccharides are generally built of simple carbohydrates called monosaccharides with general formula $(CH_2O)_n$ where *n* is three or more. A typical monosaccharide has the structure H-$(CHOH)_x$(C=O)-$(CHOH)_y$-H, that is, an aldehyde or ketone with many hydroxyl groups added, usually one on each carbon atom that is not part of the aldehyde or ketone functional group. Examples of monosaccharides are glucose, fructose, and glyceraldehyde. However, some biological substances commonly called "monosaccharides" do not conform to this formula (e.g., uronic acids and deoxy-sugars such as fucose), and there are many chemicals that do conform to this formula but are not considered to be monosaccharides (e.g., formaldehyde CH_2O and inositol $(CH_2O)_6$). The open-chain form of a monosaccharide often coexists with a closed ring form where the aldehyde/ketone carbonyl group carbon (C=O) and hydroxyl group (-OH) react forming a hemiacetal with a new C-O-C bridge. Monosaccharides can be linked together into what are called polysaccharides (or oligosaccharides) in a large variety of ways. Many carbohydrates contain one or more modified monosaccharide units that have had one or more groups replaced or removed. For example, deoxyribose, a component of DNA, is a modified version of ribose; chitin is composed of repeating units of N-acetylglucosamine, a nitrogen-containing form of glucose.

Monosaccharides

Monosaccharides are the most basic units of biologically important carbohydrates. They are the simplest form of sugar and are usually colorless, water-soluble, crystalline solids. Some monosaccharides have a sweet taste. Examples of monosaccharides include glucose (dextrose), fructose (levulose), galactose, xylose and ribose. Monosaccharides are the building blocks of disaccharides such as sucrose and polysaccharides (such as cellulose and starch). Further, each carbon atom that supports a hydroxyl group (except for the first and last) is chiral, giving rise to a number of isomeric forms all with the same chemical formula. For instance, galactose and glucose are both aldohexoses, but have different chemical and physical properties.

Structure

With few exceptions (e.g., deoxyribose), monosaccharides have the chemical formula $C_x(H_2O)_y$ with the chemical structure $H(CHOH)_nC{=}O(CHOH)_mH$. If n or m is zero, it is an aldehyde and is termed an aldose; otherwise, it is a ketone and is termed a ketose. Monosaccharides contain either a ketone or aldehyde functional group, and hydroxyl groups on most or all of the non-carbonyl carbon atoms.

Monosaccharide Momenclature

Monosaccharides are classified by the number of carbon atoms they contain:

- Triose, 3 carbon atoms
- Tetrose, 4 carbon atoms
- Pentose, 5 carbon atoms
- Hexose, 6 carbon atoms
- Heptose, 7 carbon atoms.

Some monosaccharides exist in equilibrium between its open chain and the cyclic hemiacetal forms:

Aldoses can form:

- Pyranose results from the intramolecular nucleophilic addition of the C5-OH group to the C1 carbonyl group, yielding a 6-membered ring.

Ketoses can form:

- Pyranose results from the intramolecular nucleophilic addition of the -OH at C6 to the C1 carbonyl group, forming a 6-membered ring.
- Furanose results from the intramolecular nucleophilic addition of the C4-OH group at C5 and to the C1 carbonyl group, yielding a 5-membered ring.

Monosaccharides are classified by the type of carbonyl group they contain:

- Aldose, -CHO (aldehyde)
- Ketose, C=O (ketone).

Isomerism

The total number of possible stereoisomers of one compound (n) is dependent on the number of stereogenic centres (c) in the molecule. The upper limit for the number of possible stereoisomers is $n = 2^c$. The only monosaccharide without an isomer is dihydroxyacetone or DHA.

Monosaccharides are classified according to their molecular configuration at the chiral carbon furthest removed from the aldehyde or ketone group. The chirality at this carbon is compared to the chirality of carbon 2 on glyceraldehyde. If it is equivalent to D-glyceraldehyde' s C2, the sugar is D; if it is equivalent to L-glyceraldehyde's C2, the sugar is L. Due to the chirality of the sugar molecules, an aqueous solution of a D or L saccharides will rotate light. D-glyceraldehyde causes polarized light to rotate clockwise (dextrorotary); L-glyceraldehyde causes polarized light to rotate counterclockwise (levorotary). Unlike glyceraldehyde, D/L designation on more complex sugars is not associated with their direction of light rotation. Since more complex sugars contain multiple chiral carbons, the direction of light rotation cannot be predicted by the chirality of the carbon that defines D/L nomenclature.

- D, configuration as in D-glyceraldehyde
- L, configuration as in L-glyceraldehyde.

All these classifications can be combined, resulting in names like *D-aldohexose* or *ketotriose*.

Derivatives

A large number of biologically important modified monosaccharides exist:

- Amino sugars such as:
  - Galactosamine
  - Glucosamine
  - Sialic acid
  - *N*-Acetylglucosamine
- Sulfosugars such as:
  - Sulfoquinovose.

Monosaccharides are the simplest carbohydrates in that they cannot be hydrolysed to smaller carbohydrates. They are aldehydes or ketones with two or more hydroxyl groups. The general chemical formula of an unmodified monosaccharide is $(C \cdot H_2O)_n$, literally a "carbon hydrate." Monosaccharides are important fuel molecules as well as building blocks for nucleic acids. The smallest monosaccharides, for which n = 3, are dihydroxyacetone and D- and L-glyceraldehyde.

Classification of Monosaccharides

Monosaccharides are classified according to three different characteristics: the placement of its carbonyl group, the number of carbon atoms it contains, and its chiral handedness. If the carbonyl group is an aldehyde, the monosaccharide is an aldose; if the carbonyl group is a ketone, the monosaccharide is a ketose. Monosaccharides with three carbon atoms are called trioses, those with four are called tetroses, five are called pentoses, six are hexoses, and so on. These two systems of classification are often combined. For example, glucose is an aldohexose (a six-carbon aldehyde), ribose is an aldopentose (a five-carbon aldehyde), and fructose is a ketohexose (a six-carbon ketone).

Each carbon atom bearing a hydroxyl group (-OH), with the exception of the first and last carbons, are asymmetric, making them stereocenters with two possible configurations each (R or S). Because of this asymmetry, a number of isomers may exist for any given monosaccharide formula. The aldohexose D-glucose, for example, has the formula $(C{\cdot}H_2O)_6$, of which all but two of its six carbons atoms are stereogenic, making D-glucose one of $2^4 = 16$ possible stereoisomers.

In the case of glyceraldehyde, an aldotriose, there is one pair of possible stereoisomers, which are enantiomers and epimers. 1,3-dihydroxyacetone, the ketose corresponding to the aldose glyceraldehyde, is a symmetric molecule with no stereocenters). The assignment of D or L is made according to the orientation of the asymmetric carbon furthest from the carbonyl group: in a standard Fischer projection if the hydroxyl group is on the right the molecule is a D sugar, otherwise it is an L sugar. The "D-" and "L-" prefixes should not be confused with "d-" or "l-", which indicate the direction that the sugar rotates plane polarized light. This usage of "d-" and "l-" is no longer followed in carbohydrate chemistry.

Ring-straight Chain Isomerism

The aldehyde or ketone group of a straight-chain monosaccharide will react reversibly with a hydroxyl group on a different carbon atom to form a hemiacetal or hemiketal, forming a heterocyclic ring with an oxygen bridge between two carbon atoms. Rings with five and six atoms are called furanose and pyranose forms, respectively, and exist in equilibrium with the straight-chain form.

During the conversion from straight-chain form to cyclic form, the carbon atom containing the carbonyl oxygen, called the anomeric carbon, becomes a stereogenic centre with two possible configurations: The oxygen atom may take a position either above or below the plane of the ring. The resulting possible pair of stereoisomers are called anomers. In the *α anomer*, the -OH substituent on the anomeric carbon rests on the opposite side (trans) of the ring from the CH_2OH side branch.

The alternative form, in which the CH_2OH substituent and the anomeric hydroxyl are on the same side (cis) of the plane of the ring, is called the *β anomer*. You can remember that the β anomer is cis by the mnemonic, "It's always better to βe up". Because the ring and straight-chain forms readily interconvert, both anomers exist in equilibrium. In a Fischer Projection, the α anomer is represented with the anomeric hydroxyl group *trans* to the CH_2OH and *cis* in the β anomer.

Use in Living Organisms

Monosaccharides are the major source of fuel for metabolism, being used both as an energy source (glucose being the most important in nature) and in biosynthesis. When monosaccharides are not immediately needed by many cells they are often converted to more space efficient forms, often polysaccharides. In many animals, including humans, this storage form is glycogen, especially in liver and muscle cells. In plants, starch is used for the same purpose.

Disaccharides

Two joined monosaccharides are called a disaccharide and these are the simplest polysaccharides. Examples include sucrose and lactose. They are composed of two monosaccharide units bound together by a covalent bond known as a glycosidic linkage formed via a dehydration reaction, resulting in the loss of a hydrogen atom from one monosaccharide and a hydroxyl group from the other. The formula of unmodified disaccharides is $C_{12}H_{22}O_{11}$. Although there are numerous kinds of disaccharides, a handful of disaccharides are particularly notable. Sucrose, pictured to the right, is the most abundant disaccharide, and the main form in which carbohydrates are transported in plants. It is composed of one D-glucose molecule and one D-fructose molecule. The systematic name for sucrose, *O*-α-D-glucopyranosyl-(1→2)-D-fructofuranoside, indicates four things:

- Its monosaccharides: glucose and fructose
- Their ring types: glucose is a pyranose, and fructose is a furanose
- How they are linked together: the oxygen on carbon number 1 (C1) of α-D-glucose is linked to the C2 of D-fructose.
- The *-oside* suffix indicates that the anomeric carbon of both monosaccharides participates in the glycosidic bond.

Lactose, a disaccharide composed of one D-galactose molecule and one D-glucose molecule, occurs naturally in mammalian milk. The systematic name for lactose is *O*-β-D-galactopyranosyl-(1→4)-D-glucopyranose. Other notable disaccharides include maltose (two D-glucoses linked α-1,4) and cellulobiose (two D-glucoses linked β-1,4). disaccharides can be classified into two types they are reducing and non-reducing disaccahrides if the functional group is present in bonding with another sugar unit it is called as reducing disaccharide.

Oligosaccharides and Polysaccharides

Oligosaccharides and polysaccharides are composed of longer chains of monosaccharide units bound together by glycosidic bonds. The distinction between the two is based upon the number of monosaccharide units present in the chain. Oligosaccharides typically contain between three and ten monosaccharide units, and polysaccharides contain greater than ten monosaccharide units. Definitions of how large a carbohydrate must be to fall into each category vary according to personal opinion. Examples of oligosaccharides include the disaccharides mentioned above, the trisaccharide raffinose and the tetrasaccharide stachyose.

Oligosaccharides are found as a common form of protein posttranslational modification. Such posttranslational modifications include the Lewis and ABO oligosaccharides responsible for blood group classifications and so of tissue incompatibilities, the alpha-Gal epitope responsible for hyperacute rejection in xenotransplantation, and O-GlcNAc modifications.

Polysaccharides represent an important class of biological polymers. Their function in living organisms is usually either structure- or storage-related. Starch (a polymer of glucose) is used as a storage polysaccharide in plants, being found in the form of both amylose and the branched amylopectin.

In animals, the structurally similar glucose polymer is the more densely branched glycogen, sometimes called 'animal starch'. Glycogen's properties allow it to be metabolized more quickly, which suits the active lives of moving animals.

Cellulose and chitin are examples of structural polysaccharides. Cellulose is used in the cell walls of plants and other organisms, and is claimed to be the most abundant organic molecule on earth. It has many uses such as a significant role in the paper and textile industries, and is used as a feedstock for the production of rayon (via the viscose process), cellulose acetate, celluloid, and nitrocellulose. Chitin has a similar structure, but has nitrogen-containing side branches, increasing its strength. It is found in arthropod exoskeletons and in the cell walls of some fungi. It also has multiple uses, including surgical threads.

Other polysaccharides include callose or laminarin, chrysolaminarin, xylan, arabinoxylan, mannan, fucoidan and galactomannan.

Carbohydrates are the main energy source for the human body. Chemically, carbohydrates are organic molecules in which carbon, hydrogen, and oxygen bond together in the ratio: $C_x(H_2O)_y$, where x and y are whole numbers that differ depending on the specific carbohydrate to which we are referring. Animals (including humans) break down carbohydrates during the process of metabolism to release energy. For example, the chemical metabolism of the sugar glucose is shown below:

$$C_6H_{12}O_6 + 6\ O_2\ 6\ CO_2 + 6\ H_2O + \text{energy}$$

Animals obtain carbohydrates by eating foods that contain them, for example potatoes, rice, breads, and so on. These carbohydrates are manufactured by plants during the process of photosynthesis. Plants harvest energy from sunlight to run the reaction just described in reverse:

$$6\ CO_2 + 6\ H_2O + \text{energy (from sunlight)}\ C_6H_{12}O_6 + 6\ O_2$$

A potato, for example, is primarily a chemical storage system containing glucose molecules manufactured during photosynthesis. In a potato, however, those glucose molecules are bound together in a long chain. As it turns out, there are two types of carbohydrates, the simple sugars and those carbohydrates that are made of long chains of sugars- the complex carbohydrates.

Simple Sugars

All carbohydrates are made up of units of sugar (also called saccharide units). Carbohydrates that contain only one sugar unit (monosaccharides) or two sugar units (disaccharides) are referred to as simple sugars. Simple sugars are sweet in taste and are broken down quickly in the body to release energy.

Two of the most common monosaccharides are glucose and fructose. Glucose is the primary form of sugar stored in the human body for energy. Fructose is the main sugar found in most fruits.

Both glucose and fructose have the same chemical formula ($C_6H_{12}O_6$); however, they have different structures, as shown (note: the carbon atoms that sit in the "corners" of the rings are not labelled):

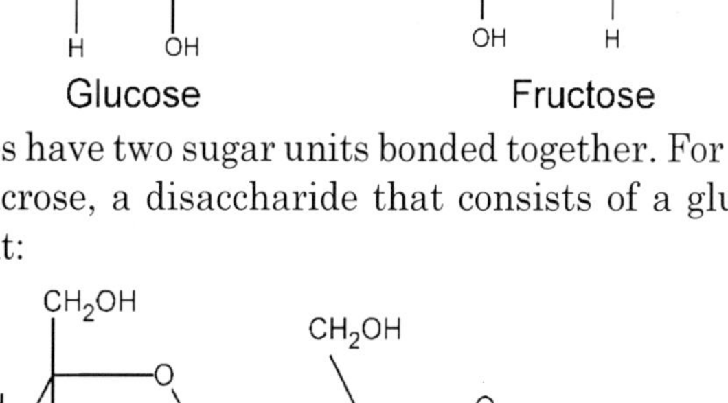

Disaccharides have two sugar units bonded together. For example, common table sugar is sucrose, a disaccharide that consists of a glucose unit bonded to a fructose unit:

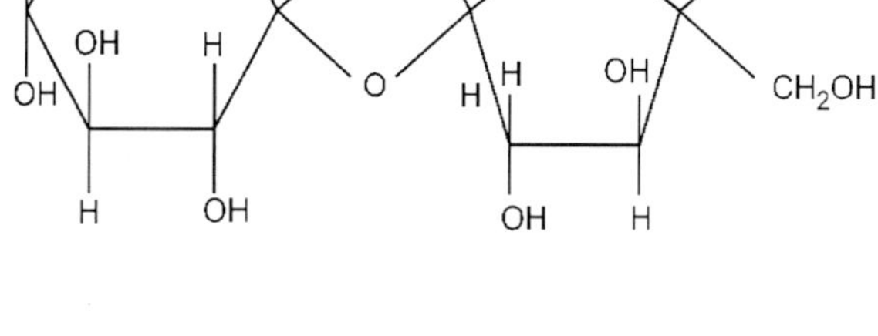

Complex carbohydrates Complex carbohydrates are polymers of the simple sugars. In other words, the complex carbohydrates are long chains of simple sugar units bonded together (for this reason the complex carbohydrates are often referred to as polysaccharides).

The potato we discussed earlier actually contains the complex carbohydrate starch. Starch is a polymer of the monosaccharide glucose:

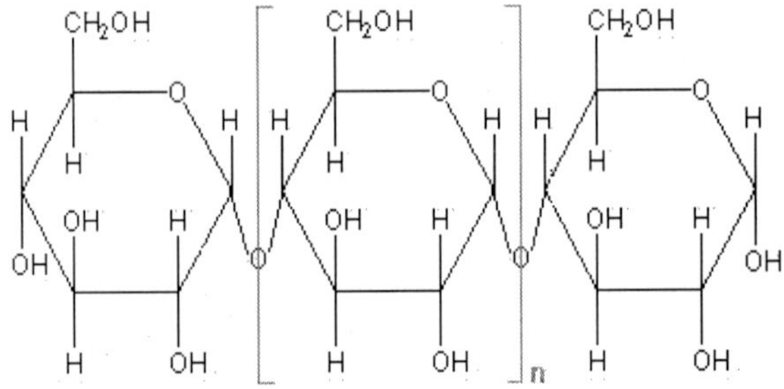

***Figure:** Starch (n is the number of repeating glucose units and ranges in the 1,000's)*

Starch is the principal polysaccharide used by plants to store glucose for later use as energy. Plants often store starch in seeds or other specialized organs; for example, common sources of starch include rice, beans, wheat, corn, potatoes, and so on. When humans eat starch, an enzyme that occurs in saliva and in the intestines called amylase breaks the bonds between the repeating glucose units, thus allowing the sugar to be absorbed into the bloodstream.

Once absorbed into the bloodstream, the human body distributes glucose to the areas where it is needed for energy or stores it as its own special polymer-glycogen. Glycogen, another polymer of glucose, is the polysaccharide used by animals to store energy.

Excess glucose is bonded together to form glycogen molecules, which the animal stores in the liver and muscle tissue as an "instant" source of energy. Both starch and glycogen are polymers of glucose; however, starch is a long, straight chain of glucose units, whereas glycogen is a branched chain of glucose units.

Another important polysaccharide is cellulose. Cellulose is yet a third polymer of the monosaccharide glucose. Cellulose differs from starch and glycogen because the glucose units form a two-dimensional structure, with hydrogen bonds holding together nearby polymers, thus giving the molecule added stability.

Cellulose, also known as plant fiber, cannot be digested by human beings, therefore cellulose passes through the digestive tract without being absorbed into the body. Some animals, such as cows and termites, contain bacteria in their digestive tract that help them to digest cellulose. Cellulose is a relatively stiff material, and in plants it is used as a structural molecule to add support to the leaves, stem, and other plant parts. Despite the fact that it cannot be used as an energy source in most animals, cellulose fiber is essential in the diet because it helps exercise the digestive track and keep it clean and healthy.

How Organisms Convert Food to Energy

Do you ever wonder how organisms convert food to energy?

We all know that organisms must take in nutrients to survive. This self-instructional unit describes two of the most important processes in food to energy conversion. These processes, called glycolysis and respiration, use the energy of the glucose molecule to manufacture ATP, the most widely used energy molecule of the cell.

ATP is an abbreviation for the nucleotide Adenosine triphosphate. This nucleotide consists of a ribose, a phosphate and the nitrogenous base adenine. This is the same nucleotide used to manufacture RNA. Don't be surprised by the fact that ATP is very energetic. All of the nucleotides used to make RNA and DNA are full of energy.

Most of their energy is stored in the two bonds between their two terminal phosphates. The nucleotides of DNA are called deoxyribonucleotides (abbreviated dATP, dTTP, dCTP and dGTP). The nucleotides of RNA are called ribonucleotides (abbreviated ATP, UTP, CTP, and GTP). While these nucleotides have different types of ribose sugar and nitrogenous bases, they all have the two terminal phosphate groups, and so they all have equivalent energies. All the energy in the nucleotide comes from the two terminal phosphates.

The energy used to make DNA and RNA polymers, through the formation of phosphodiester bonds, is supplied by the triphosphate nucleotides when they polymerize to form nucleic acids. It takes energy to form all chemical bonds, and the breaking of these chemical bonds will release energy.

The triphosphate nucleotides are special because the bonds between the phosphates take more energy to form (~8 kilocalories/mole) as compared to the bonds between the phosphate and the ribose (~5 kilocalories/mole). Therefore these high energy bonds between the phosphates release more energy when they are broken.

The phosphates in the structural formula of ATP have been assigned specific labels based on the IUPAC system. The phosphate attached to the #5 carbon of the ribose is labelled as the alpha (a) phosphate. The next two phosphates are labelled beta (ß) and gamma (g) from left to right. The bonds between the beta and gamma phosphates are represented by wavy lines which indicate the higher energy potential of these bonds.

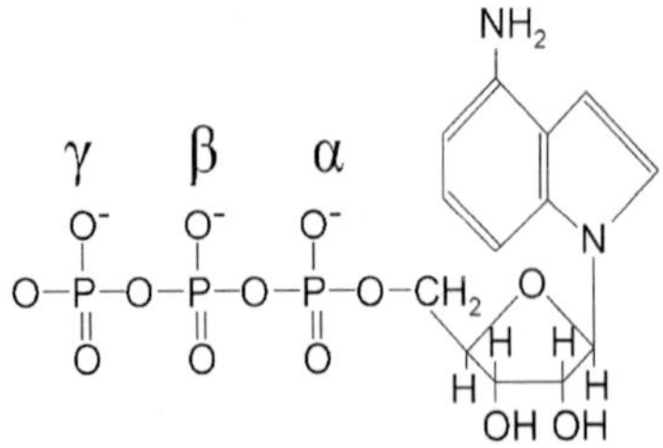

Figure: *The structure of ATP*

Most biochemical reactions require energy input to proceed. These types of reactions are called endergonic reactions. An exergonic reaction, by contrast, gives off energy and proceeds spontaneously. Many chemical reactions of the cell use the energy from ATP which released when the beta or gamma phosphate bonds of ATP are broken. The energy released from the ATP supplies the energy necessary to form or break chemical bonds in biochemical reactions.

The energy released by the conversion of ATP to ADP is used in the manufacture of ATP during the process of glycolysis. When the phosphates are cleaved from ATP, two other forms of the nucleotide are made; ADP (adenosine diphosphate) and AMP (adenosine monophosphate). If you cover the terminal (gamma) phosphate in the structural formula of ATP with a piece of paper you will see what ADP looks like and if you cover the gamma and beta phosphates you will see what AMP looks like.

ATP is made inside cells by the transfer of a phosphate on to the b phosphate of ADP. Where does ADP come from? ADP is always inside cells as the byproduct of the cleavage of a gamma phosphate from ATP. This is the ultimate use of recycling by the cell! Sometimes the ATP, ADP and AMP molecules are completely degraded because they wear out. Cells contain many biochemical pathways and can rebuild new AMP molecules from scratch when necessary. The AMP molecule can then be phosphorylated to form ADP and ATP. However, the cell prefers to conserve energy, whenever possible, by the dephosphorylation of ATP to ADP and AMP and the rephosphorylation of these back to ATP. The following reaction represents the phosphorylation and dephosphorylation reactions of ATP.

$$\text{ATP} \rightleftharpoons \text{ADP} \rightleftharpoons \text{AMP}$$

How Cells Use Glucose for the Manufacture of ATP

The first step of the breakdown of glucose inside the cell, to utilize the potential energy stored in this molecule, is a process called glycolysis. Any basic college biology text will have all of the steps of the process of glycolysis. In this self instructional I want to give you an overview of the process, in the hope that you will understand what is taking place and not just memorize the process.

Where does glycolysis take place?

Glycolysis takes place inside the cytoplasm of the cell. The cytoplasm is the aqueous based solution inside a cell that has a variety of molecules solubilized in it. The contents of the cytoplasm include proteins, carbohydrates, nucleic acids, salts and a whole host of other soluble molecules.

What are the parts of the cell that are not considered cytoplasm?

The organelles inside the cytoplasm are not part of the cytoplasm. Remember, most organelles are bound by lipid membranes and are not soluble in the cytoplasm. The contents of organelles are also isolated from the cytoplasm due to the hydrophobic nature of their membranes. This means that the substances solubilized inside organelles stay separate from the rest of the cytoplasm. The enzymes necessary for glycolysis are solubilized in the cytoplasm of the cell.

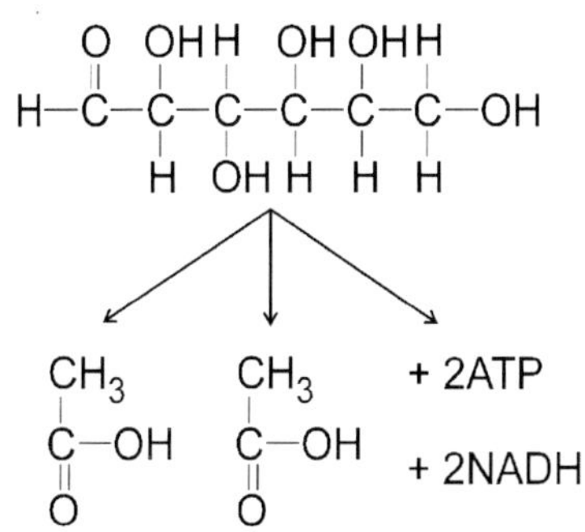

Figure: *Splitting of glucose through glycolysis*

What does the process of glycolysis accomplish?

- Glycolysis uses the energy from two ATP to perform a series of enzymatic reactions resulting in the manufacture of four ATP.
- Glycolysis also generates two three-carbon compounds called pyruvate that still contain energy. The destination for pyruvate is the Citric Acid Cycle (also known as Krebs cycle) which occurs inside the matrix of the mitochondria.
- Glycolysis results in the addition of two hydrogens (electrons) onto the energy carrier nicotinamide adenine dinucleotide (NAD) to make NADH. The process of adding electrons to a substrate is called reduction. The destination for the NADH is the electron transport chain also found in the mitochondria.

Why are only two net ATP produced during glycolysis?

Although glycolysis allows for the formation of four ATP, two ATP are used in the process. This results in a net profit of two ATP from the process of Glycolysis.

Does glycolysis require molecular oxygen? No, glycolysis is an anaerobic process.

Enzymes and Enzymatic Reactions

To help understand the processes of glucose manufacture, anabolic reactions, and catabolic reactions, we need to understand how enzymes work.

Enzymes are a group of biological molecules, most of which are made of proteins. Enzymes are biological catalysts. A catalyst is added to a chemical reaction to speed the rate of the reaction. In other words, a catalyst makes the reaction occur faster.

The catalyst is not changed in its form, but instead retains its original physical and chemical characteristics at the completion of the reaction. A catalyst does not participate in a reaction and can be reused again and again. Enzymes have all the qualities of a catalyst, but they are specifically used in the chemical reactions of living organisms (metabolism). Often people say; enzymes break this substance down or form this substance; however, to state the function of the enzyme correctly, one should say enzymes catalyse the formation or breakdown of a substance.

How do enzymes work? Enzymatic catalysts reduce the amount of energy necessary for endergonic reactions to proceed. Enzymes don't supply the energy, they reduce the amount needed. How? Well, there are several theories concerning this. One way to speed reactions is to concentrate the materials that participate in them. Remember the generalized formula for chemical reactions?

Reactant A + Reactant B ———> Product C + Product D

Enzymes are molecules that have a 3-dimensional shape. Think of a muffin tin with all the depressions used to hold the individual muffins. If one hole contains reactant A and the adjacent hole contains reactant B, the two reactants are in close proximity to one another. This has the effect of concentrating the two reactants necessary for an anabolic reaction to occur and form products C and D.

Remember the cytoplasm of the cell or even the contents of an organelle may seem tiny, but it is a vast ocean of many components to a glucose or other small molecule.

Concentrating reactants is one way to speed reactions. Another way to reduce the energy and speed the rate of catabolic reactions might be to hold a molecule in place so that some of its bonds are under strain. Imagine you were trying to cut a thick piece of cloth with dull scissors and the going was very slow. If two people were to hold the material taut, or under strain, the fabric would cut more easily. Perhaps enzymes hold molecules in a similar fashion allowing bonds to be broken more easily.

If enzymes only speed the rates of reactions, does this mean that chemical reactions occur spontaneously anyway? Yes, all chemical reactions can occur spontaneously, but some reactions may take millennia to occur. Since organisms do not live on a geological time scale, the ability of enzymes to speed the rate of reactions, up to thousands of times faster, is necessary for life to exist.

The Structure of Enzymes

Enzymes are 3-dimensional molecules that have folded up into a specific shape (eg. the muffin tin). Imagine the depression in our muffin tin to be the site at which the chemical reaction takes place. This site is called the active site of the enzyme. The molecule that fits in the active site of the enzyme is called the substrate. Enzymes and substrates are very specific for each other. The active site of an enzyme is designed to fit perfectly with its specific substrate, like a 3-dimensional jigsaw puzzle.

Biochemical or Enzymatic Pathways

Sometimes a cell needs a product from a reactant that requires only a single chemical reaction. For example, reactant A needs to be changed to product B. This reaction would only require the enzyme that fits with its substrate, reactant A. However, this is not usually the case. Most products inside cells that need to be broken down or built up require a multi-step process. This process is called a biochemical pathway. It goes something like this:

- Enzyme A changes substrate A to product B.
- Product or substrate B now can be bound by enzyme B and be changed to product or substrate C
- Substrate C now fits with enzyme C which changes it to product D.
- If product D is what the cell needs, the pathway is finished. If not, the pathway can continue in the same way until the ultimate product is obtained.

Finally a quick lesson in naming enzymes. Generally to name an enzyme we take the substrate it breaks down or product that it manufactures and add an -ase on the end.

Examples: enzymes that degrade the following substances DNAse, ATPase or enzymes that build up the following substances: DNA polymerase and ATP synthetase.

Glucose Utilization

The figure below demonstrates some of the possible pathways for glucose utilization by a variety of organisms.

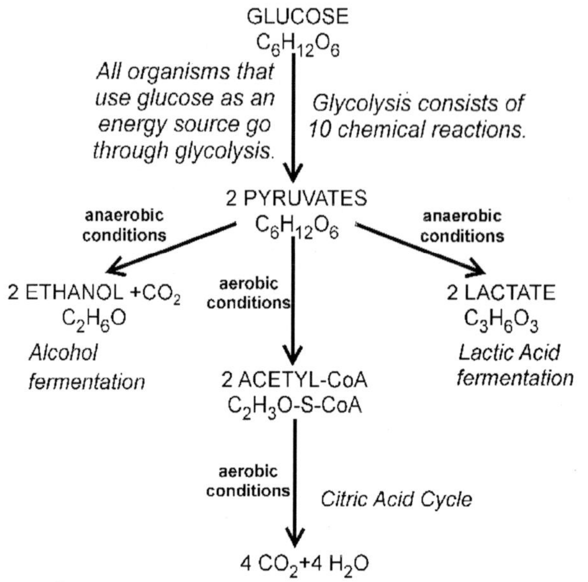

Figure: *Pyruvate metabolism in different organisms*

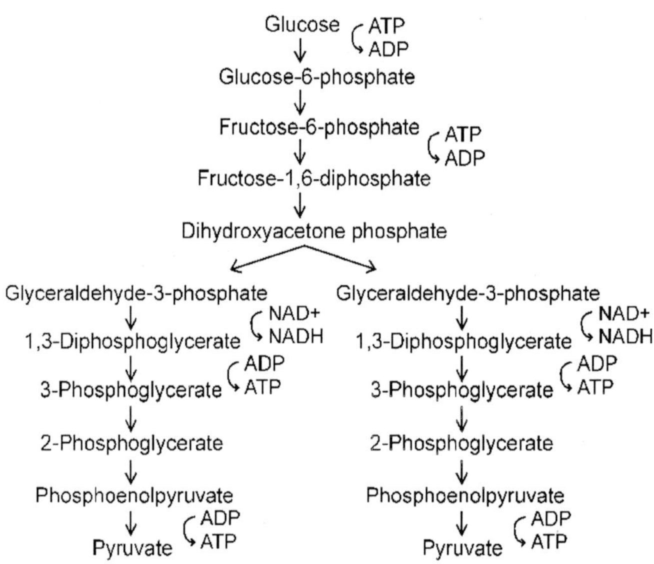

Figure: *Glycolysis as a Biochemical Pathway*

Please note that the conversion of glucose to pyruvate requires 10 enzyme reactions and occurs in 10 separate steps. After glycolysis, what happens to the pyruvate molecules?

Both of the pyruvate (three-carbon molecules) are immediately converted to two-carbon molecules by removing a CO_2 (carbon dioxide) from each molecule. The remaining two-carbon compound is known as an acetyl group. The acetyl group will react quickly with another molecule because of its unpaired electrons. Therefore as the CO_2 is cleaved from pyruvate, the acetyl group is added simultaneously to a carrier molecule called coenzyme A (CoA).

Coenzyme A (CoA) is not an enzyme but a molecule similar in some respects to a nucleotide. This is a very important carrier molecule because it carries the acetyl group to many important biochemical pathways. The acetyl group and CoA molecule combine to form acetyl CoA. This molecule travels to the matrix of the mitochondria.

Once there, the CoA transfers the two-carbon acetyl group to an enzyme which catalyzes the addition of a four-carbon compound called oxaloacetic acid (oxaloacetate) to the acetyl. The result of this reaction is a six-carbon compound called citric acid (citrate). This is why this biochemical pathway is called the citric acid cycle.

The citric acid molecule has carbon, hydrogen, and oxygen atoms stripped from it in a step-wise fashion through this biochemical pathway. Each step requires its own specific enzyme. The end result of this pathway is the regeneration of oxaloacetic acid, and the conversion of NAD and FAD to NADH and FADH with the formation of carbon dioxide CO_2 as a byproduct.

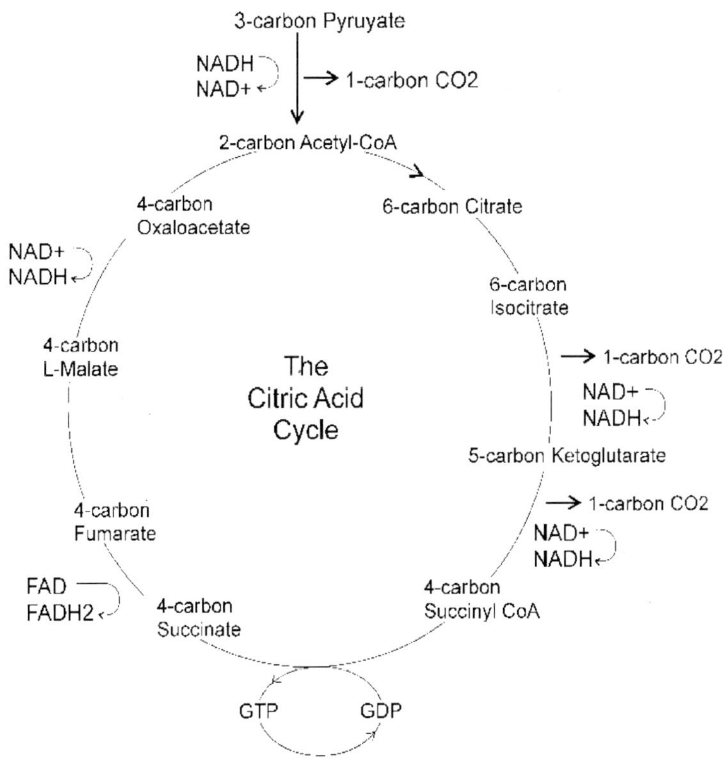

Figure: *A simplified diagram of the citric acid (Krebs) cycle.*

The major objective of the citric acid cycle is to get the hydrogens, along with their electrons, off the carbon compounds so they can be carried to the electron transport proteins that are embedded in the membrane of the mitochondrion.

The waste or by-products of these reactions are the carbon and oxygen that are stripped off as each compound is rearranged by the removal of hydrogen. These carbons and oxygens combine to form CO_2 carbon dioxide gas which is given off as a byproduct of these reactions. The CO_2 that you breathe out of your lungs is a waste product coming from the cellular metabolism of glucose! Carbon dioxide (CO_2) enters our bloodstream, eventually winds up in our lungs, and is expelled upon exhalation, but where do those hydrogens with their electrons go? The hydrogens and electrons are picked up by electron carrier NAD (nicotinamide adenine dinucleotide) and electron carrier FAD (flavin adenine dinucleotide).

These electron carriers transfer the hydrogens and electrons to the electron carrying proteins, embedded in the mitochondrial membrane, which form the electron transport chain. The electron transport chain proteins are complexed with metal ions which allow them to carry extra electrons safely.

When NADH and FADH arrive at the electron transport chain, they transfer the electrons associated with the hydrogens to the electron carrying proteins at the top of the electron chain.

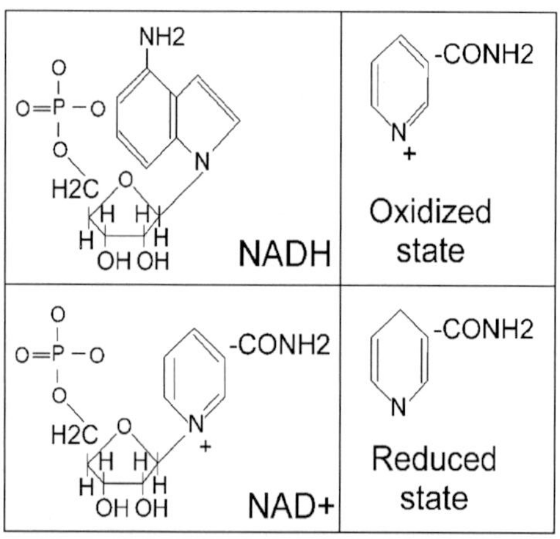

Figure: *The structural formula for NAD and NADH.*

When electrons and hydrogen atoms are added to a molecule, this is called reduction. When electrons or hydrogens are removed from a molecule, this is known as oxidation. NAD is in an oxidized state while NADH is in a reduced state. Now that the electrons have been transferred to the electron transport proteins, the hydrogen ion H+ (proton) is free to be pumped across the membrane into a space between the inner and outer mitochondrial membranes.

This proton pumping requires energy which is generated by the flow of electrons as they are passed down the electron transport protein chain. As electrons are transferred from one protein to the next, energy is released at each transfer, until the final electron acceptor, oxygen is reached. Aerobic organisms must take in oxygen because it serves as the final electron acceptor in the electron transport chain. The energy, released from the passage of the electrons down the electron transport chain to oxygen, is used to pump the protons across the membrane. The pumping of protons requires a great deal of energy, because they have to be pumped against a concentration gradient.

Because protons are constantly being pumped across the membrane, they fill the space between the mitochondrial membranes. This is known as an area of high proton concentration. Putting more protons in this area would be like trying to force balloons into a box already filled to capacity with balloons.

Pushing protons against a concentration gradient of protons is referred to as going against, or up, a concentration gradient. In the case of the balloons, as soon as you open the box they would come out easily with no energy input necessary.

This is termed going with, or down, the concentration gradient. Molecules naturally flow from areas of high concentration to areas of low concentration with no energy input necessary.

Once you have captured all this energy by pushing protons up the proton concentration gradient into a crowded area, what happens? The protons can get out but they only have one passage to escape. This is a passage created by a protein enzyme called ATP synthetase. The ATP synthetase phosphorylates (adds phosphates to) ADP. The ATP synthetase gets the energy necessary for the formation of the high energy phosphate bond from the rush of protons through the channel it creates across the mitochondrial membrane.

To better understand how the flow of protons provides energy for the ATP synthetase enzyme imagine how a turbine engine works at Hoover Dam. The rushing water is caught and causes the huge turbine to turn which generates electricity. In a similar way the energy of protons cascading through the ATP synthetase is converted into the energy used to form the phosphate bond of ATP. This process, called oxidative phosphorylation, makes most of the ATP manufactured by cells. While glycolysis gives us net profit of two ATP, the electron transport coupled with the citric acid cycle and oxidative phosphorylation give us a net profit of approximately 34 ATP from each glucose catabolized. Quite a big difference in yield!

What happens to the protons and electrons when they flow back through the membrane and down electron transport chain? Since electrons are energetic enough to cause problems inside the cell they must be inactivated. The final electron acceptor, oxygen, combines with the electrons and excess hydrogen ions to form water, a non-toxic by-product. So the products of respiration are CO_2 (from the citric acid cycle) and H_2O (from electron transport) and lots of ATP. Without oxygen, aerobic cells die because the manufacture of ATP is halted at the cellular level. Since oxygen is absolutely necessary for the continual phosphorylation of ADP to ATP this process is called oxidative phosphorylation. Without oxygen, electron transport would stop, protons would not be pumped, and the ATP synthetase would not have a flow of protons to supply the energy necessary for the manufacture of ATP.

Do you know where all the molecules come from and their destination in the following Generalized Equation for Respiration?

$$C6H_{12}O_6 + O_2 \longrightarrow CO_2 + H_2O + \text{energy}$$

Now I would like to give you my overview of all of the processes of glucose metabolism in narrative form. I hope that you will understand the processes of glycolysis and respiration well enough to be able to describe this process in a similar way. In other words, I would like you to be able to think through and understand the concepts, not the minute details of these processes.

- Glucose is taken up as food by the body and carried to a cell. Glucose enters the cell by facilitated diffusion and is immediately broken down by the process of glycolysis in the cytoplasm of the cell.
- The process of glycolysis uses a biochemical pathway consisting of 10 enzyme mediated reactions which uses two ATP and manufactures four ATP (a net profit of two ATP). Glycolysis generates two NADH, which carry hydrogen and electrons to the electron transport chain in the mitochondrial membrane. Glycolysis also generates two three-carbon pyruvate molecules.
- The pyruvate molecule has a CO_2 removed and the remaining two-carbon acetyl group is then added onto the carrier molecule Coenzyme A to form acetyl CoA.
- Acetyl CoA is taken to the mitochondrial matrix to participate in the Citric acid cycle. The two-carbon acetyl group is then added via an enzyme mediated reaction to a four-carbon oxaloacetic acid molecule which results in the formation of the six-carbon molecule citric acid. Through a series of enzyme mediated reactions oxygens, carbons and hydrogens are systematically stripped off. The carbons and oxygens combine in the appropriate proportions to form CO2 and ultimately are expelled as waste products.
- The hydrogen atoms along with their electrons are picked up by the electron carriers NAD and FAD and taken to the electron transport proteins, which are located on the inner mitochondrial membrane.

- The electrons are handed off to the electron transport proteins where they flow from one protein to another releasing energy at each transfer until they combine with oxygen, the final electron acceptor in the electron transport chain.
- The energy released from the flow of electrons down the transport chain is used to pump protons across the inner mitochondrial membrane into the space between the inner and outer mitochondrial membranes, against a proton concentration gradient.
- The protons then flow back down the concentration gradient through an ATP synthetase enzyme channel, back into the mitochondrial matrix. The flow of protons through the ATP synthetase generates the energy necessary to phosphorylate ADP to ATP.
- Once back inside the matrix the protons combine with oxygen and electrons to form water a by-product of oxidative phosphorylation.

Protein Structure and Function

Proteins (also known as polypeptides) are organic compounds made of amino acids arranged in a linear chain and folded into a globular form. The amino acids in a polymer are joined together by the peptide bonds between the carboxyl and amino groups of adjacent amino acid residues. The sequence of amino acids in a protein is defined by the sequence of a gene, which is encoded in the genetic code. In general, the genetic code specifies 20 standard amino acids; however, in certain organisms the genetic code can include selenocysteine—and in certain archaea—pyrrolysine. Shortly after or even during synthesis, the residues in a protein are often chemically modified by post-translational modification, which alters the physical and chemical properties, folding, stability, activity, and ultimately, the function of the proteins. Proteins can also work together to achieve a particular function, and they often associate to form stable complexes.

Of the most distinguishing features of polypeptides is their ability to fold into a globular state, or "structure". The extent to which proteins fold into a defined structure varies widely. Data supports that some protein structures fold into a highly rigid structure with small fluctuations and are therefore considered to be single structure. Other proteins have been shown to undergo large rearrangements from one conformation to another. This conformational change is often associated with a signalling event. Thus, the structure of a protein serves as a medium through which to regulate either the function of a protein or activity of an enzyme. Not all proteins requiring a folding process in order to function as some function in an unfolded state.

Like other biological macromolecules such as polysaccharides and nucleic acids, proteins are essential parts of organisms and participate in virtually every process within cells. Many proteins are enzymes that catalyse biochemical reactions and are vital to metabolism. Proteins also have structural or mechanical functions, such as actin and myosin in muscle and the proteins in the cytoskeleton, which form a system of scaffolding that maintains cell shape. Other proteins are important in cell signalling, immune responses, cell adhesion, and the cell cycle. Proteins are also necessary in animals' diets, since animals cannot synthesize all the amino acids they need and must obtain essential amino acids from food. Through the process of digestion, animals break down ingested protein into free amino acids that are then used in metabolism.

Proteins were first described by the Dutch chemist Gerhardus Johannes Mulder and named by the Swedish chemist Jöns Jakob Berzelius in 1838. Early nutritional scientists such as the German Carl von Voit believed that protein was the most important nutrient for maintaining the structure of the body, because it was generally believed that "flesh makes flesh." The central role of proteins as enzymes in living organisms was however not fully appreciated until 1926, when James B. Sumner showed that the enzyme urease was in fact a protein. The first protein to be sequenced was insulin, by Frederick Sanger, who won the Nobel Prize for this achievement in 1958. The first protein structures to be solved were hemoglobin and myoglobin, by Max Perutz and Sir John Cowdery Kendrew, respectively, in 1958. The three-dimensional structures of both proteins were first determined by x-ray diffraction analysis; Perutz and Kendrew shared the 1962 Nobel Prize in Chemistry for these discoveries. Proteins may be purified from other cellular components using a variety of techniques such as ultracentrifugation, precipitation, electrophoresis, and chromatography; the advent of genetic engineering has made possible a number of methods to facilitate purification. Methods commonly used to study protein structure and function include immunohistochemistry, site-directed mutagenesis, nuclear magnetic resonance and mass spectrometry.

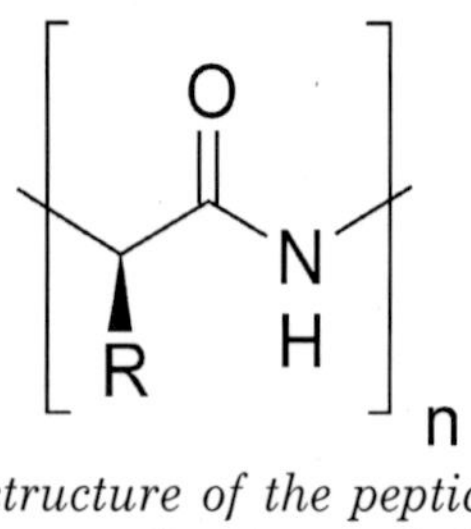

Figure: *Chemical structure of the peptide bond (left) and a peptide bond between leucine and threonine (right).*

Most proteins are linear polymers built from series of up to 20 different L-α-amino acids. All amino acids possess common structural features, including an α-carbon to which an amino group, a carboxyl group, and a variable side chain are bonded. Only proline differs from this basic structure as it contains an unusual ring to the N-end amine group, which forces the CO–NH amide moiety into a fixed conformation. The side chains of the standard amino acids, detailed in the list of standard amino acids, have a great variety of chemical structures and properties; it is the combined effect of all of the amino acid side chains in a protein that ultimately determines its three-dimensional structure and its chemical reactivity.

The amino acids in a polypeptide chain are linked by peptide bonds. Once linked in the protein chain, an individual amino acid is called a *residue,* and the linked series of carbon, nitrogen, and oxygen atoms are known as the *main chain* or *protein backbone.* The peptide bond has two resonance forms that contribute some double-bond character and inhibit rotation around its axis, so that the alpha carbons are roughly coplanar. The other two dihedral angles in the peptide bond determine the local shape assumed by the protein backbone. The end of the protein with a free carboxyl group is known as the C-terminus or carboxy terminus, whereas the end with a free amino group is known as the N-terminus or amino terminus.

The words *protein, polypeptide,* and *peptide* are a little ambiguous and can overlap in meaning. *Protein* is generally used to refer to the complete biological molecule in a stable conformation, whereas *peptide* is generally reserved for a short amino acid oligomers often lacking a stable three-dimensional structure. However, the boundary between the two is not well defined and usually lies near 20–30 residues. *Polypeptide* can refer to any single linear chain of amino acids, usually regardless of length, but often implies an absence of a defined conformation.

Synthesis

Proteins are assembled from amino acids using information encoded in genes. Each protein has its own unique amino acid sequence that is specified by the nucleotide sequence of the gene encoding this protein. The genetic code is a set of three-nucleotide sets called codons and each three-nucleotide combination designates an amino acid, for example AUG (adenine-uracil-guanine) is the code for methionine. Because DNA contains four nucleotides, the total number of possible codons is 64; hence, there is some redundancy in the genetic code, with some amino acids specified by more than one codon.

Genes encoded in DNA are first transcribed into pre-messenger RNA (mRNA) by proteins such as RNA polymerase. Most organisms then process the pre-mRNA (also known as a *primary transcript*) using various forms of post-transcriptional modification to form the mature mRNA, which is then used as a template for protein synthesis by the ribosome. In prokaryotes the mRNA may either be used as soon as it is produced, or be bound by a ribosome after having moved away from the nucleoid. In contrast, eukaryotes make mRNA in the cell nucleus and then translocate it across the nuclear membrane into the cytoplasm, where protein synthesis then takes place. The rate of protein synthesis is higher in prokaryotes than eukaryotes and can reach up to 20 amino acids per second.

The process of synthesizing a protein from an mRNA template is known as translation. The mRNA is loaded onto the ribosome and is read three nucleotides at a time by matching each codon to its base pairing anticodon located on a transfer RNA molecule, which carries the amino acid corresponding to the codon it recognizes. The enzyme aminoacyl tRNA synthetase "charges" the tRNA molecules with the correct amino acids. The growing polypeptide is often termed the *nascent chain*. Proteins are always biosynthesized from N-terminus to C-terminus. The size of a synthesized protein can be measured by the number of amino acids it contains and by its total molecular mass, which is normally reported in units of *daltons* (synonymous with atomic mass units), or the derivative unit kilodalton (kDa). Yeast proteins are on average 466 amino acids long and 53 kDa in mass. The largest known proteins are the titins, a component of the muscle sarcomere, with a molecular mass of almost 3,000 kDa and a total length of almost 27,000 amino acids.

Chemical Synthesis

Short proteins can also be synthesized chemically by a family of methods known as peptide synthesis, which rely on organic synthesis techniques such as chemical ligation to produce peptides in high yield. Chemical synthesis allows for the introduction of non-natural amino acids into polypeptide chains, such as attachment of fluorescent probes to amino acid side chains.

These methods are useful in laboratory biochemistry and cell biology, though generally not for commercial applications. Chemical synthesis is inefficient for polypeptides longer than about 300 amino acids, and the synthesized proteins may not readily assume their native tertiary structure. Most chemical synthesis methods proceed from C-terminus to N-terminus, opposite the biological reaction.

Structure

Proteins are an important class of biological macromolecules present in all organisms. All proteins are polymers of amino acids. Classified by their physical size, proteins are nanoparticles (definition: 1-100 nm). Each protein polymer – also known as a polypeptide – consists of a sequence of 20 different L-α-amino acids, also referred to as residues. For chains under 40 residues the term peptide is frequently used instead of protein.

To be able to perform their biological function, proteins fold into one or more specific spatial conformations, driven by a number of non-covalent interactions such as hydrogen bonding, ionic interactions, Van Der Waals forces, and hydrophobic packing. To understand the functions of proteins at a molecular level, it is often necessary to determine their three-dimensional structure. This is the topic of the scientific field of structural biology, which employs techniques such as X-ray crystallography, NMR spectroscopy, and dual polarisation interferometry to determine the structure of proteins.

Protein structures range in size from tens to several thousand residues Very large aggregates can be formed from protein subunits: for example, many thousand actin molecules assemble into a microfilament.

A protein may undergo reversible structural changes in performing its biological function. The alternative structures of the same protein are referred to as different conformations, and transitions between them are called conformational changes.

Protein Covalent Structure and Stereochemistry

Protein amino acids are combined into a single polypeptide chain in a condensation reaction. This reaction is catalysed by the ribosome in a process known as translation.

Amino Acid Residues

Each α-amino acid consists of a backbone part that is present in all the amino acid types, and a side chain that is unique to each type of residue. An exception from this rule is proline, where the hydrogen atom is replaced by a bond to the side chain. Because the carbon atom is bound to four different groups it is chiral, however only one of the isomers occur in biological proteins. Glycine however, is not chiral since its side chain is a hydrogen atom. A simple mnemonic for correct L-form is "CORN": when the C_α atom is viewed with the H in front, the residues read "CO-R-N" in a clockwise direction.

The 20 naturally occurring amino acids have different physical and chemical properties, including their electrostatic charge, pKa, hydrophobicity, size and specific functional groups. These properties play a major role in molding protein structure.

The Peptide Bond

The peptide bond tend to be planar due to the delocalization of the electrons from the double bond. The rigid peptide dihedral angle, ω (the bond between C_1 and N) is always close to 180 degrees. The dihedral angles phi φ (the bond between N and Cα) and psi ø (the bond between Cα and C_1) can have a certain range of possible values.

These angles are the internal degrees of freedom of a protein, they control the protein's conformation. They are restrained by geometry to allowed ranges typical for particular secondary structure elements, and represented in a Ramachandran plot.

Side-chain Conformation

The atoms along the side chain are named with Greek letters in Greek alphabetical order: α, β, γ, δ, °, and so on. C_α refers to the carbon atom of the backbone closest to the carbonyl group of that amino acid, C_β the second closest and so on. The dihedral angles around the bonds between these atoms are named ÷1, ÷2, ÷3, etc. The dihedral angle of the first movable atom of the side chain, γ, defined as N-Cα-Cβ-*X*γ, is named ÷1. Side chains tend to adopt different staggered conformations called *gauche(-)*, *trans*, and *gauche(+)*, which corresponds to rotation angles of 60°, 180°, and -60°, respectively, around the sp3-sp3 bonds.

The diversity of side-chain conformations is often expressed in rotamer libraries. A rotamer library is a collection of rotamers for each residue type. Side-chain dihedral angles are not evenly distributed, but for most side chain types, the ÷ angles occur in tight clusters around certain values. Rotamer libraries therefore are usually derived from statistical analysis of side-chain conformations in known structures of proteins by clustering observed conformations or by dividing dihedral angle space into bins, and determining an average conformation in each bin·

Levels of Protein Structure

There are four distinct levels of protein structure.

Primary Structure

The primary structure refers to the sequence of the different amino acids of the peptide or protein. The primary structure is held together by covalent or peptide bonds, which are made during the process of protein biosynthesis or translation. The two ends of the polypeptide chain are referred to as the carboxyl terminus (C-terminus) and the amino terminus (N-terminus) based on the nature of the free group on each extremity. Counting of residues always starts at the N-terminal end (NH_2-group), which is the end where the amino group is involved in a peptide bond.

The primary structure of a protein is determined by the gene corresponding to the protein. A specific sequence of nucleotides in DNA is transcribed into mRNA, which is read by the ribosome in a process called translation. The sequence of a protein is unique to that protein, and defines the structure and function of the protein.

The sequence of a protein can be determined by methods such as Edman degradation or tandem mass spectrometry. Often however, it is read directly from the sequence of the gene using the genetic code. Post-translational modifications such as disulfide formation, phosphorylations and glycosylations are usually also considered a part of the primary structure, and cannot be read from the gene.

Secondary Structure

Secondary structure can be formally defined by the hydrogen bonds of the biopolymer, as observed in an atomic-resolution structure. In proteins, the secondary structure is defined by the patterns of hydrogen bonds between backbone amide and carboxyl groups. In nucleic acids, the secondary structure is defined by the hydrogen bonding between the nitrogenous bases. The hydrogen bonding patterns may be significantly distorted, which makes an automatic determination of secondary structure difficult.

The secondary structure may be also defined based on the regular pattern of backbone dihedral angles in a particular region of the Ramachandran plot; thus, a segment of residues with such dihedral angles may be called a helix, regardless of whether it has the correct hydrogen bonds. The secondary structure may be also provided by crystallographers in the corresponding PDB file.

The rough secondary-structure content of a biopolymer (e.g., "this protein is 40% α-helix and 20% β-sheet.") can often be estimated spectroscopically. For proteins, a common method is far-ultraviolet (far-UV, 170-250 nm) circular dichroism. A pronounced double minimum at 208 and 222 nm indicate α-helical structure, whereas a single minimum at 204 nm or 217 nm reflects random-coil or β-sheet structure, respectively. A less common method is infrared spectroscopy, which detects differences in the bond oscillations of amide groups due to hydrogen-bonding. Finally, secondary-structure contents may be estimated accurately using the chemical shifts of an unassigned NMR spectrum.

Protein

Secondary structure in proteins consists of local inter-residue interactions mediated by hydrogen bonds, or not. The most common secondary structures are alpha helices and beta sheets. Other helices, such as the 3_{10} helix and ð helix, are calculated to have energetically favourable hydrogen-bonding patterns but are rarely if ever observed in natural proteins except at the ends of α helices due to unfavourable backbone packing in the centre of the helix. Other extended structures such as the polyproline helix and alpha sheet are rare in native state proteins but are often hypothesized as important protein folding intermediates. Tight turns and loose, flexible loops link the more "regular" secondary structure elements. The random coil is not a true secondary structure, but is the class of conformations that indicate an absence of regular secondary structure.

Amino acids vary in their ability to form the various secondary structure elements. Proline and glycine are sometimes known as "helix breakers" because they disrupt the regularity of the α helical backbone conformation; however, both have unusual conformational abilities and are commonly found in turns. Amino acids that prefer to adopt helical conformations in proteins include methionine, alanine, leucine, glutamate and lysine ("MALEK" in amino-acid 1-letter codes); by contrast, the large aromatic residues (tryptophan, tyrosine and phenylalanine) and C^{β}-branched amino acids (isoleucine, valine, and threonine) prefer to adopt β-strand conformations. However, these preferences are not strong enough to produce a reliable method of predicting secondary structure from sequence alone.

There are several methods for defining protein secondary structure (e.g. DEFINE, DSSP, STRIDE (protein)).

The DSSP Code

The Dictionary of Protein Secondary Structure, in short DSSP, is commonly used to describe the protein secondary structure with single letter codes. The secondary structure is assigned based on hydrogen bonding patterns as those initially proposed by Pauling et al. in 1951 (before any protein structure had ever been experimentally determined). There are eight types of secondary structure that DSSP defines:

- G = 3-turn helix (3_{10} helix). Min length 3 residues.
- H = 4-turn helix (α helix). Min length 4 residues.
- I = 5-turn helix (ð helix). Min length 5 residues.
- T = hydrogen bonded turn (3, 4 or 5 turn)
- E = extended strand in parallel and/or anti-parallel b-sheet conformation. Min length 2 residues.
- B = residue in isolated b-bridge (single pair b-sheet hydrogen bond formation)
- S = bend (the only non-hydrogen-bond based assignment).

Amino acid residues which are not in any of the above conformations are assigned as the eighth type 'Coil': often codified as ' ' (space), C (coil) or '-' (dash). The helices (G,H and I) and sheet conformations are all required to have a reasonable length.

This means that 2 adjacent residues in the primary structure must form the same hydrogen bonding pattern. If the helix or sheet hydrogen bonding pattern is too short they are designated as T or B, respectively. Other protein secondary structure assignment categories exist (sharp turns, Omega loops etc.), but they are less frequently used.

DSSP H-bond Definition

Secondary structure is defined by hydrogen bonding, so the exact definition of a hydrogen bond is critical. The standard H-bond definition for secondary structure is that of DSSP, which is a purely electrostatic model. It assigns charges of $\pm q_1 \equiv 0.42e$ to the carbonyl carbon and oxygen, respectively, and charges of $\pm q_2 \equiv 0.20e$ to the amide nitrogen and hydrogen, respectively. The electrostatic energy is:

$$E = q_1 q_2 \left[\frac{1}{r_{ON}} + \frac{1}{r_{CH}} - \frac{1}{r_{OH}} - \frac{1}{r_{CN}} \right] .332\,\text{kcal/mol.}$$

According to DSSP, an H-bond exists if and only if E is less than -0.5 kcal/mol. Although the DSSP formula is a relatively crude approximation of the *physical* H-bond energy, it is generally accepted as a tool for defining secondary structure.

Protein Secondary-structure Prediction

Predicting protein tertiary structure from only its amino acid sequence is a very challenging problem, but using the simpler secondary structure definitions is more tractable and has been the focus for research for a long time.

Although, the 8-state DSSP code is already a simplification from the continuous variation of hydrogen bonding patterns present in a protein the majority of secondary prediction methods simplify further to the three dominant states: Helix, Sheet and Coil. How the conversion is made from 8- to 3-state varies between methods.

Early methods of secondary-structure prediction were based on the helix- or sheet-forming propensities of individual amino acids, sometimes coupled with rules for estimating the free energy of forming secondary structure elements. Such methods were typically ~60% accurate in predicting which of the three states (helix/sheet/coil) a residue adopts. A significant increase in accuracy (to nearly ~80%) was made by exploiting multiple sequence alignment; knowing the full distribution of amino acids that occur at a position (and in its vicinity, typically ~7 residues on either side) throughout evolution provides a much better picture of the structural tendencies near that position.

For illustration, a given protein might have a glycine at a given position, which by itself might suggest a random coil there. However, multiple sequence alignment might reveal that helix-favouring amino acids occur at that position (and nearby positions) in 95% of homologous proteins spanning nearly a billion years of evolution.

Moreover, by examining the average hydrophobicity at that and nearby positions, the same alignment might also suggest a pattern of residue solvent accessibility consistent with an α-helix. Taken together, these factors would suggest that the glycine of the original protein adopts α-helical structure, rather than random coil. Several types of methods are used to combine all the available data to form a 3-state prediction, including neural networks, hidden Markov models and support vector machines. Modern prediction methods also provide a confidence score for their predictions at every position. Secondary-structure prediction methods are continuously benchmarked, e.g., in the EVA experiment. Based on ~270 weeks of testing, the most accurate methods at present are PSIPRED, SAM, PORTER, PROF and SABLE.

Interestingly, it does not seem to be possible to improve upon these methods by taking a consensus of them· The chief area for improvement appears to be the prediction of β-strands; residues confidently predicted as β-strand are likely to be so, but the methods are apt to overlook some β-strand segments (false negatives). There is likely an upper limit of ~90% prediction accuracy overall, due to the idiosyncrasies of the standard method (DSSP) for assigning secondary-structure classes (helix/strand/coil) to PDB structures, against which the predictions are benchmarked·

Accurate secondary-structure prediction is a key element in the prediction of tertiary structure, in all but the simplest (homology modeling) cases.

Alignment

Both protein and nucleic acid secondary structures can be used to aid in multiple sequence alignment. These alignments can be made more accurate by the inclusion of secondary structure information in addition to simple sequence information. This is sometimes less useful in RNA because base pairing is much more highly conserved than sequence. Distant relationships between proteins whose primary structures are unalignable can sometimes be found by secondary structure.

Tertiary Structure

Tertiary structure refers to three-dimensional structure of a single protein molecule. The alpha-helices and beta-sheets are folded into a compact globule. The folding is driven by the *non-specific* hydrophobic interactions (the burial of hydrophobic residues from water), but the structure is stable only when the parts of a protein domain are locked into place by *specific* tertiary interactions, such as salt bridges, hydrogen bonds, and the tight packing of side chains and disulfide bonds. The disulfide bonds are extremely rare in cytosolic proteins, since the cytosol is generally a reducing environment.

Relationship to Primary Structure

Tertiary structure is considered to be largely determined by the protein's primary structure, or the sequence of amino acids of which it is composed. Efforts to predict tertiary structure from the primary structure are known generally as protein structure prediction.

However, the environment in which a protein is synthesized and allowed to fold are significant determinants of its final shape and are usually not directly taken into account by current prediction methods. (Most such methods do rely on comparisons between the sequence to be predicted and sequences of known structure in the Protein Data Bank and thus account for environment indirectly, assuming the target and template sequences share similar cellular contexts.)

Determinants of Tertiary Structure

In globular proteins, tertiary interactions are frequently stabilized by the sequestration of hydrophobic amino acid residues in the protein core, from which water is excluded, and by the consequent enrichment of charged or hydrophilic residues on the protein's water-exposed surface.

In secreted proteins that do not spend time in the cytoplasm, disulfide bonds between cysteine residues help to maintain the protein's tertiary structure. A variety of common and stable tertiary structures appear in a large number of proteins that are unrelated in both function and evolution- for example, many proteins are shaped like a TIM barrel, named for the enzyme triosephosphateisomerase. Another common structure is a highly stable dimeric coiled coil structure composed of 2-7 alpha helices. Proteins are classified by the folds they represent in databases like SCOP and CATH.

Stability of Native States

The most typical conformation of a protein in its cellular environment is generally referred to as the native state or native conformation. It is commonly assumed that this most-populated state is also the most thermodynamically stable conformation attainable for a given primary structure; this is a reasonable first approximation but the claim assumes that the reaction is not under kinetic control- that is, that the time required for the protein to attain its native conformation before being translated is small.

In the cell, a variety of protein chaperones assist a newly synthesized polypeptide in attaining its native conformation. Some such proteins are highly specific in their function, such as protein disulfide isomerase; others are very general and can be of assistance to most globular proteins- the prokaryotic GroEL/GroES system and the homologous eukaryotic Heat shock proteins Hsp60/Hsp10 system fall into this category.

Some proteins explicitly take advantage of the fact that they can become kinetically trapped in a relatively high-energy conformation due to folding kinetics. Influenza hemagglutinin, for example, is synthesized as a single polypeptide chain that acts as a kinetic trap.

The "mature" activated protein is proteolytically cleaved to form two polypeptide chains that are trapped in a high-energy conformation. Upon encountering a drop in pH, the protein undergoes an energetically favourable conformational rearrangement that enables it to penetrate a host cell membrane. Many serpins (serine protease inhibitors) are metastable, and undergo a conformational change when a loop of the protein is cut by a protease.

Quaternary Structure

Quaternary structure is a larger assembly of several protein molecules or polypeptide chains, usually called subunits in this context. The quaternary structure is stabilized by the same non-covalent interactions and disulfide bonds as the tertiary structure. Complexes of two or more polypeptides (i.e. multiple subunits) are called multimers. Specifically it would be called a dimer if it contains two subunits, a trimer if it contains three subunits, and a tetramer if it contains four subunits. The subunits are frequently related to one another by symmetry operations, such as a 2-fold axis in a dimer. Multimers made up of identical subunits are referred to with a prefix of "homo-" (e.g. a homotetramer) and those made up of different subunits are referred to with a prefix of "hetero-" (e.g. a heterotetramer, such as the two alpha and two beta chains of hemoglobin). Many proteins do not have the quaternary structure and function as monomers.

Many proteins are actually assemblies of more than one polypeptide chain, which in the context of the larger assemblage are known as protein subunits. In addition to the tertiary structure of the subunits, multiple-subunit proteins possess a quaternary structure, which is the arrangement into which the subunits assemble. Enzymes composed of subunits with diverse functions are sometimes called holoenzymes, in which some parts may be known as regulatory subunits and the functional core is known as the catalytic subunit. Examples of proteins with quaternary structure include hemoglobin, DNA polymerase, and ion channels. Other assemblies referred to instead as *multiprotein complexes* also possess quaternary structure. Examples include nucleosomes and microtubules. Changes in quaternary structure can occur through conformational changes within individual subunits or through reorientation of the subunits relative to each other. It is through such changes, which underlie cooperativity and allostery in "multimeric" enzymes, that many proteins undergo regulation and perform their physiological function.

The above definition follows a classical approach to biochemistry, established at times when the distinction between a protein and a functional, proteinaceous unit was difficult to elucidate. More recently, people refer to protein-protein interaction when discussing quaternary structure of proteins and consider all assemblies of proteins as protein complexes.

Nomenclature of Quaternary Structures

The number of subunits in an oligomeric complex are described using names that end in -mer. Formal and Greco-Latinate names are generally used for the first ten types and can be used for up to twenty subunits, whereas higher order complexes are usually described by the number of subunits, followed by -meric.

- 1 = monomer
- 2 = dimer
- 3 = trimer
- 4 = tetramer
- 5 = pentamer
- 6 = hexamer
- 7 = heptamer
- 8 = octamer
- 9 = nonamer
- 10 = decamer
- 11 = undecamer
- 12 = dodecamer
- 13 = tridecamer
- 14 = tetradecamer
- 15 = pentadecamer*
- 16 = hexadecamer
- 17 = heptadecamer*
- 18 = octadecamer
- 19 = nonadecamer
- 20 = eicosamer
- 21-mer
- 22-mer
- 23-mer* etc.

**No known examples*

Although complexes higher than octamers are rarely observed for most proteins, there are some important exceptions. Viral capsids are often composed of multiples of 60 proteins.

Several molecular machines are also found in the cell, such as the proteasome (four heptameric rings = 28 subunits), the transcription complex and the spliceosome. The ribosome is probably the largest molecular machine, and is composed of many RNA and protein molecules.

In some cases, proteins form complexes that then assemble into even larger complexes. In such cases, one uses the nomenclature, e.g., "dimer of dimers" or "trimer of dimers", to suggest that the complex might dissociate into smaller sub-complexes before dissociating into monomers.

Determination of Quaternary Structure

Protein quaternary structure can be determined using a variety of experimental techniques that require a sample of protein in a variety of experimental conditions. The experiments often provide an estimate of the mass of the native protein and, together with knowledge of the masses and/or stoichiometry of the subunits, allow the quaternary structure to be predicted with a given accuracy. It is not always possible to obtain a precise determination of the subunit composition for a variety of reasons. The number of subunits in a protein complex can often be determined by measuring the hydrodynamic molecular volume or mass of the intact complex, which requires native solution conditions. For *folded* proteins, the mass can be inferred from its volume using the partial specific volume of 0.73 ml/g. However, volume measurements are less certain than mass measurements, since *unfolded* proteins appear to have a much larger volume than folded proteins; additional experiments are required to determined whether a protein is unfolded or has formed an oligomer.

Prediction of Quaternary Structure Attribute

Some bioinformatics methods were developed for predicting the quaternary structural attributes of proteins based on their sequence information by using various modes of pseudo amino acid composition.

Methods that measure mass of intact complex directly:

- sedimentation-equilibrium analytical ultracentrifugation
- electrospray mass spectrometry
- Mass spectrometric immunoassay (MSIA)

Methods that measure the size of the intact complex directly:

- static light scattering
- size exclusion chromatography (requires calibration)
- Dual polarisation interferometry

Methods that measure the size of the intact complex indirectly:

- sedimentation-velocity analytical ultracentrifugation (measures the translational diffusion constant)
- dynamic light scattering (measures the translational diffusion constant)
- pulsed-gradient protein nuclear magnetic resonance (measures the translational diffusion constant)
- fluorescence polarization (measures the rotational diffusion constant)
- dielectric relaxation (measures the rotational diffusion constant)
- Dual polarisation interferometry (measures the size and the density of the complex)

Methods that measure the mass or volume under unfolding conditions (such as MALDI-TOF mass spectrometry and SDS-PAGE) are generally not useful, since non-native conditions usually cause the complex to dissociate into monomers. However, these may sometimes be applicable; for example, the experimenter may apply SDS-PAGE after first treating the intact complex with chemical cross-linking reagents.

Protein-protein Interactions

Proteins are capable of forming very tight complexes. For example, ribonuclease inhibitor binds to ribonuclease A with a roughly 20 fM dissociation constant. Other proteins have evolved to bind specifically to unusual moieties on another protein, e.g., biotin groups (avidin), phosphorylated tyrosines (SH2 domains) or proline-rich segments (SH3 domains).

Domains, Motifs, and Folds in Protein Structure

Protein are frequently described as consisting from several structural units. A structural domain is an element of the protein's overall structure that is self-stabilizing and often folds independently of the rest of the protein chain. Many domains are not unique to the protein products of one gene or one gene family but instead appear in a variety of proteins. Domains often are named and singled out because they figure prominently in the biological function of the protein they belong to; for example, the "calcium-binding domain of calmodulin". Because they are independently stable, domains can be "swapped" by genetic engineering between one protein and another to make chimeras.

The structural and sequence motifs refer to short segments of protein three-dimensional structure or amino acid sequence that were found in a large number of different proteins.

The supersecondary structure refers to a specific combination of secondary structure elements, such as beta-alpha-beta units or helix-turn-helix motif. Some of them may be also referred to as structural motifs.

Protein fold refers to the general protein architecture, like helix bundle, beta-barrel, Rossman fold or different "folds" provided in the Structural Classification of Proteins database. Despite the fact that there are about 100,000 different proteins expressed in eukaryotic systems, there are many fewer different domains, structural motifs and folds. This is partly a consequence of evolution, since genes or parts of genes can be doubled or moved around within the genome.

This means that, for example, a protein domain might be moved from one protein to another thus giving the protein a new function. Because of these mechanisms, pathways and mechanisms tend to be reused in several different proteins.

Protein Folding

An unfolded polypeptide folds into its characteristic three-dimensional structure from random coil.

Protein Structure Determination

Around 90% of the protein structures available in the Protein Data Bank have been determined by X-ray crystallography. This method allows one to measure the 3D density distribution of electrons in the protein (in the crystallized state) and thereby infer the 3D coordinates of all the atoms to be determined to a certain resolution.

Roughly 9% of the known protein structures have been obtained by Nuclear Magnetic Resonance techniques. The secondary structure composition can be determined via circular dichroism or dual polarisation interferometry. Cryo-electron microscopy has recently become a means of determining protein structures to high resolution (less than 5 angstroms or 0.5 nanometer) and is anticipated to increase in power as a tool for high resolution work in the next decade. This technique is still a valuable resource for researchers working with very large protein complexes such as virus coat proteins and amyloid fibers.

Structure Classification

Protein structures can be classified based on their similarity or a common evolutionary origin. SCOP and CATH databases provide two different structural classifications of proteins.

Computational Prediction of Protein Structure

The generation of a protein sequence is much easier than the determination of a protein structure. However, the structure of a protein gives much more insight in the function of the protein than its sequence. Therefore, a number of methods for the computational prediction of protein structure from its sequence have been developed· *Ab initio* prediction methods use just the sequence of the protein.

Threading and Homology Modeling methods can build a 3D model for a protein of unknown structure from experimental structures of evolutionary related proteins. Most proteins fold into unique 3-dimensional structures. The shape into which a protein naturally folds is known as its native conformation. Although many proteins can fold unassisted, simply through the chemical properties of their amino acids, others require the aid of molecular chaperones to fold into their native states. Biochemists often refer to four distinct aspects of a protein's structure:

- *Primary structure*: the amino acid sequence.

- *Secondary structure*: regularly repeating local structures stabilized by hydrogen bonds. The most common examples are the alpha helix, beta sheet and turns. Because secondary structures are local, many regions of different secondary structure can be present in the same protein molecule.
- *Tertiary structure*: the overall shape of a single protein molecule; the spatial relationship of the secondary structures to one another. Tertiary structure is generally stabilized by nonlocal interactions, most commonly the formation of a hydrophobic core, but also through salt bridges, hydrogen bonds, disulfide bonds, and even post-translational modifications. The term "tertiary structure" is often used as synonymous with the term *fold*. The tertiary structure is what controls the basic function of the protein.
- *Quaternary structure*: the structure formed by several protein molecules (polypeptide chains), usually called *protein subunits* in this context, which function as a single protein complex.

Proteins are not entirely rigid molecules. In addition to these levels of structure, proteins may shift between several related structures while they perform their functions. In the context of these functional rearrangements, these tertiary or quaternary structures are usually referred to as "conformations", and transitions between them are called *conformational changes*. Such changes are often induced by the binding of a substrate molecule to an enzyme's active site, or the physical region of the protein that participates in chemical catalysis. In solution proteins also undergo variation in structure through thermal vibration and the collision with other molecules.

Proteins can be informally divided into three main classes, which correlate with typical tertiary structures: globular proteins, fibrous proteins, and membrane proteins. Almost all globular proteins are soluble and many are enzymes. Fibrous proteins are often structural, such as collagen, the major component of connective tissue, or keratin, the protein component of hair and nails. Membrane proteins often serve as receptors or provide channels for polar or charged molecules to pass through the cell membrane.

A special case of intramolecular hydrogen bonds within proteins, poorly shielded from water attack and hence promoting their own dehydration, are called dehydrons.

Structure Determination

Discovering the tertiary structure of a protein, or the quaternary structure of its complexes, can provide important clues about how the protein performs its function. Common experimental methods of structure determination include X-ray crystallography and NMR spectroscopy, both of which can produce information at atomic resolution. However, NMR experiments are able to provide information from which a subset of distances between pairs of atoms can be estimated, and the final possible conformations for a protein are determined by solving a distance geometry problem. Dual polarisation interferometry is a quantitative analytical method for measuring the overall protein conformation and conformational changes due to interactions or other stimulus. Circular dichroism is another laboratory technique for determining internal beta sheet/helical composition of proteins.

Cryoelectron microscopy is used to produce lower-resolution structural information about very large protein complexes, including assembled viruses; a variant known as electron crystallography can also produce high-resolution information in some cases, especially for two-dimensional crystals of membrane proteins. Solved structures are usually deposited in the Protein Data Bank (PDB), a freely available resource from which structural data about thousands of proteins can be obtained in the form of Cartesian coordinates for each atom in the protein.

Many more gene sequences are known than protein structures. Further, the set of solved structures is biased toward proteins that can be easily subjected to the conditions required in X-ray crystallography, one of the major structure determination methods. In particular, globular proteins are comparatively easy to crystallize in preparation for X-ray crystallography. Membrane proteins, by contrast, are difficult to crystallize and are underrepresented in the PDB. Structural genomics initiatives have attempted to remedy these deficiencies by systematically solving representative structures of major fold classes. Protein structure prediction methods attempt to provide a means of generating a plausible structure for proteins whose structures have not been experimentally determined.

Cellular Functions

Proteins are the chief actors within the cell, said to be carrying out the duties specified by the information encoded in genes. With the exception of certain types of RNA, most other biological molecules are relatively inert elements upon which proteins act. Proteins make up half the dry weight of an *Escherichia coli* cell, whereas other macromolecules such as DNA and RNA make up only 3% and 20%, respectively. The set of proteins expressed in a particular cell or cell type is known as its proteome.

The chief characteristic of proteins that also allows their diverse set of functions is their ability to bind other molecules specifically and tightly. The region of the protein responsible for binding another molecule is known as the binding site and is often a depression or "pocket" on the molecular surface. This binding ability is mediated by the tertiary structure of the protein, which defines the binding site pocket, and by the chemical properties of the surrounding amino acids' side chains. Protein binding can be extraordinarily tight and specific; for example, the ribonuclease inhibitor protein binds to human angiogenin with a sub-femtomolar dissociation constant ($<10^{-5}$ M) but does not bind at all to its amphibian homolog onconase (>1 M). Extremely minor chemical changes such as the addition of a single methyl group to a binding partner can sometimes suffice to nearly eliminate binding; for example, the aminoacyl tRNA synthetase specific to the amino acid valine discriminates against the very similar side chain of the amino acid isoleucine.

Proteins can bind to other proteins as well as to small-molecule substrates. When proteins bind specifically to other copies of the same molecule, they can oligomerize to form fibrils; this process occurs often in structural proteins that consist of globular monomers that self-associate to form rigid fibers. Protein–protein interactions also regulate enzymatic activity, control progression through the cell cycle, and allow the assembly of large protein complexes that carry out many closely related reactions with a common biological function. Proteins can also bind to, or even be integrated into, cell membranes. The ability of binding partners to induce conformational changes in proteins allows the construction of enormously complex signalling networks. Importantly, as interactions between proteins are reversible, and depend heavily on the availability of different groups of partner proteins to form aggregates that are capable to carry out discrete sets of function, study of the interactions between specific proteins is a key to understand important aspects of cellular function, and ultimately the properties that distinguish particular cell types.

Enzymes

The best-known role of proteins in the cell is as enzymes, which catalyse chemical reactions. Enzymes are usually highly specific and accelerate only one or a few chemical reactions. Enzymes carry out most of the reactions involved in metabolism, as well as manipulating DNA in processes such as DNA replication, DNA repair, and transcription. Some enzymes act on other proteins to add or remove chemical groups in a process known as post-translational modification. About 4,000 reactions are known to be catalysed by enzymes. The rate acceleration conferred by enzymatic catalysis is often enormous—as much as 10^{5}-fold increase in rate over the uncatalyzed reaction in the case of orotate decarboxylase (78 million years without the enzyme, 18 milliseconds with the enzyme).

The molecules bound and acted upon by enzymes are called substrates. Although enzymes can consist of hundreds of amino acids, it is usually only a small fraction of the residues that come in contact with the substrate, and an even smaller fraction—three to four residues on average—that are directly involved in catalysis. The region of the enzyme that binds the substrate and contains the catalytic residues is known as the active site.

Cell Signalling and Ligand Binding

Many proteins are involved in the process of cell signalling and signal transduction. Some proteins, such as insulin, are extracellular proteins that transmit a signal from the cell in which they were synthesized to other cells in distant tissues. Others are membrane proteins that act as receptors whose main function is to bind a signalling molecule and induce a biochemical response in the cell. Many receptors have a binding site exposed on the cell surface and an effector domain within the cell, which may have enzymatic activity or may undergo a conformational change detected by other proteins within the cell.

Antibodies are protein components of adaptive immune system whose main function is to bind antigens, or foreign substances in the body, and target them for destruction. Antibodies can be secreted into the extracellular environment or anchored in the membranes of specialized B cells known as plasma cells. Whereas enzymes are limited in their binding affinity for their substrates by the necessity of conducting their reaction, antibodies have no such constraints. An antibody's binding affinity to its target is extraordinarily high.

Many ligand transport proteins bind particular small biomolecules and transport them to other locations in the body of a multicellular organism. These proteins must have a high binding affinity when their ligand is present in high concentrations, but must also release the ligand when it is present at low concentrations in the target tissues. The canonical example of a ligand-binding protein is haemoglobin, which transports oxygen from the lungs to other organs and tissues in all vertebrates and has close homologs in every biological kingdom. Lectins are sugar-binding proteins which are highly specific for their sugar moieties. Lectins typically play a role in biological recognition phenomena involving cells and proteins. Receptors and hormones are highly specific binding proteins.

Transmembrane proteins can also serve as ligand transport proteins that alter the permeability of the cell membrane to small molecules and ions. The membrane alone has a hydrophobic core through which polar or charged molecules cannot diffuse. Membrane proteins contain internal channels that allow such molecules to enter and exit the cell. Many ion channel proteins are specialized to select for only a particular ion; for example, potassium and sodium channels often discriminate for only one of the two ions.

Structural Proteins

Structural proteins confer stiffness and rigidity to otherwise-fluid biological components. Most structural proteins are fibrous proteins; for example, actin and tubulin are globular and soluble as monomers, but polymerize to form long, stiff fibers that comprise the cytoskeleton, which allows the cell to maintain its shape and size. Collagen and elastin are critical components of connective tissue such as cartilage, and keratin is found in hard or filamentous structures such as hair, nails, feathers, hooves, and some animal shells. Other proteins that serve structural functions are motor proteins such as myosin, kinesin, and dynein, which are capable of generating mechanical forces. These proteins are crucial for cellular motility of single celled organisms and the sperm of many multicellular organisms which reproduce sexually. They also generate the forces exerted by contracting muscles.

Cellular Localization

The study of proteins *in vivo* is often concerned with the synthesis and localization of the protein within the cell. Although many intracellular proteins are synthesized in the cytoplasm and membrane-bound or secreted proteins in the endoplasmic reticulum, the specifics of how proteins are targeted to specific organelles or cellular structures is often unclear. A useful technique for assessing cellular localization uses genetic engineering to express in a cell a fusion protein or chimera consisting of the natural protein of interest linked to a "reporter" such as green fluorescent protein (GFP). The fused protein's position within the cell can be cleanly and efficiently visualized using microscopy, as shown in the figure opposite.

Other methods for elucidating the cellular location of proteins requires the use of known compartmental markers for regions such as the ER, the Golgi, lysosomes/vacuoles, mitochondria, chloroplasts, plasma membrane, etc. With the use of fluorescently tagged versions of these markers or of antibodies to known markers, it becomes much simpler to identify the localization of a protein of interest. For example, indirect immunofluorescence will allow for fluorescence colocalization and demonstration of location. Fluorescent dyes are used to label cellular compartments for a similar purpose.

Other possibilities exist, as well. For example, immunohistochemistry usually utilizes an antibody to one or more proteins of interest that are conjugated to enzymes yielding either luminescent or chromogenic signals that can be compared between samples, allowing for localization information. Another applicable technique is cofractionation in sucrose (or other material) gradients using isopycnic centrifugation. While this technique does not prove colocalization of a compartment of known density and the protein of interest, it does increase the likelihood, and is more amenable to large-scale studies.

Finally, the gold-standard method of cellular localization is immunoelectron microscopy. This technique also uses an antibody to the protein of interest, along with classical electron microscopy techniques. The sample is prepared for normal electron microscopic examination, and then treated with an antibody to the protein of interest that is conjugated to an extremely electro-dense material, usually gold. This allows for the localization of both ultrastructural details as well as the protein of interest. Through another genetic engineering application known as site-directed mutagenesis, researchers can alter the protein sequence and hence its structure, cellular localization, and susceptibility to regulation. This technique even allows the incorporation of unnatural amino acids into proteins, using modified tRNAs, and may allow the rational design of new proteins with novel properties.

Proteomics and Bioinformatics

The total complement of proteins present at a time in a cell or cell type is known as its proteome, and the study of such large-scale data sets defines the field of proteomics, named by analogy to the related field of genomics. Key experimental techniques in proteomics include 2D electrophoresis, which allows the separation of a large number of proteins, mass spectrometry, which allows rapid high-throughput identification of proteins and sequencing of peptides (most often after in-gel digestion), protein microarrays, which allow the detection of the relative levels of a large number of proteins present in a cell, and two-hybrid screening, which allows the systematic exploration of protein–protein interactions.

The total complement of biologically possible such interactions is known as the interactome. A systematic attempt to determine the structures of proteins representing every possible fold is known as structural genomics.

The large amount of genomic and proteomic data available for a variety of organisms, including the human genome, allows researchers to efficiently identify homologous proteins in distantly related organisms by sequence alignment. Sequence profiling tools can perform more specific sequence manipulations such as restriction enzyme maps, open reading frame analyses for nucleotide sequences, and secondary structure prediction. From this data phylogenetic trees can be constructed and evolutionary hypotheses developed using special software like ClustalW regarding the ancestry of modern organisms and the genes they express. The field of bioinformatics seeks to assemble, annotate, and analyze genomic and proteomic data, applying computational techniques to biological problems such as gene finding and cladistics.

Structure Prediction and Simulation

Complementary to the field of structural genomics, protein structure prediction seeks to develop efficient ways to provide plausible models for proteins whose structures have not yet been determined experimentally.

The most successful type of structure prediction, known as homology modeling, relies on the existence of a "template" structure with sequence similarity to the protein being modeled; structural genomics' goal is to provide sufficient representation in solved structures to model most of those that remain. Although producing accurate models remains a challenge when only distantly related template structures are available, it has been suggested that sequence alignment is the bottleneck in this process, as quite accurate models can be produced if a "perfect" sequence alignment is known.

Many structure prediction methods have served to inform the emerging field of protein engineering, in which novel protein folds have already been designed. A more complex computational problem is the prediction of intermolecular interactions, such as in molecular docking and protein–protein interaction prediction.

The processes of protein folding and binding can be simulated using such technique as molecular mechanics, in particular, molecular dynamics and Monte Carlo, which increasingly take advantage of parallel and distributed computing (Folding@Home project; molecular modeling on GPU).

The folding of small alpha-helical protein domains such as the villin headpiece and the HIV accessory protein have been successfully simulated *in silico*, and hybrid methods that combine standard molecular dynamics with quantum mechanics calculations have allowed exploration of the electronic states of rhodopsins.

Nutrition

Most microorganisms and plants can biosynthesize all 20 standard amino acids, while animals (including humans) must obtain some of the amino acids from the diet. The amino acids that an organism cannot synthesize on its own are referred to as essential amino acids. Key enzymes that synthesize certain amino acids are not present in animals — such as aspartokinase, which catalyzes the first step in the synthesis of lysine, methionine, and threonine from aspartate.

If amino acids are present in the environment, microorganisms can conserve energy by taking up the amino acids from their surroundings and downregulating their biosynthetic pathways. In animals, amino acids are obtained through the consumption of foods containing protein. Ingested proteins are then broken down into amino acids through digestion, which typically involves denaturation of the protein through exposure to acid and hydrolysis by enzymes called proteases.

Some ingested amino acids are used for protein biosynthesis, while others are converted to glucose through gluconeogenesis, or fed into the citric acid cycle.

This use of protein as a fuel is particularly important under starvation conditions as it allows the body's own proteins to be used to support life, particularly those found in muscle. Amino acids are also an important dietary source of nitrogen.

Peptide Bond

A peptide bond (amide bond) is a covalent chemical bond formed between two molecules when the carboxyl group of one molecule reacts with the amino group of the other molecule, thereby releasing a molecule of water (H_2O).

This is a dehydration synthesis reaction (also known as a condensation reaction), and usually occurs between amino acids. The resulting C(O)NH bond is called a peptide bond, and the resulting molecule is an amide. The four-atom functional group -C(=O)NH- is called a peptide link. Polypeptides and proteins are chains of amino acids held together by peptide bonds, as is the backbone of PNA.

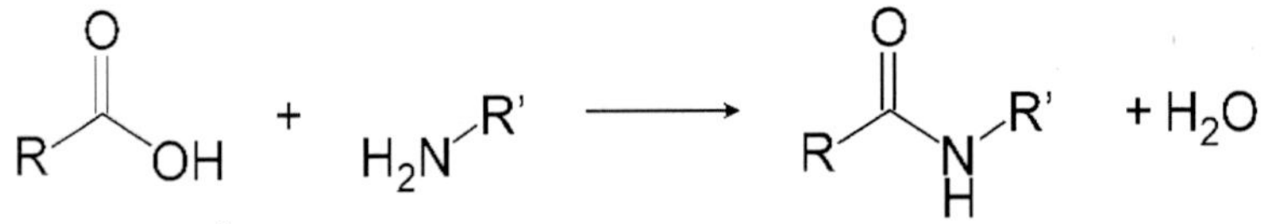

Figure: *Dehydration synthesis (condensation) reaction forming an amide*

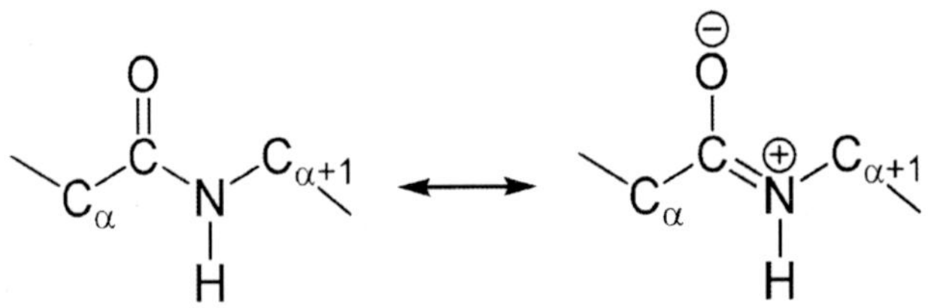

Figure: *Resonance forms of a typical peptide group. The uncharged, single-bonded form (typically ~60%) is shown on the left, whereas the charged, double-bonded form (typically ~40%) is on the right.*

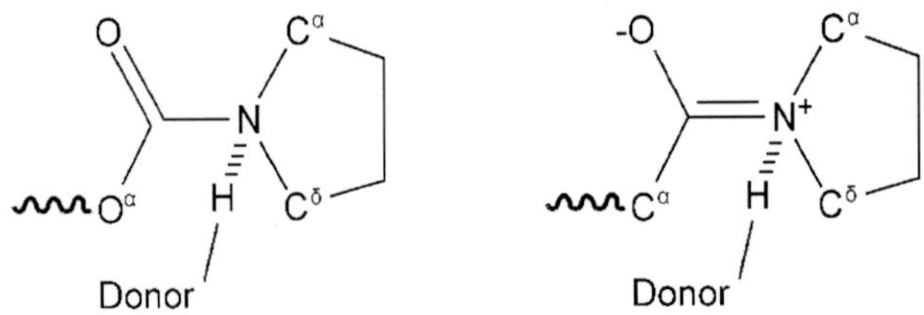

Figure: *A hydrogen bond (dotted line) to an X-Pro peptide group favors the single-bonded resonance form (left) over the double-bonded form (right).*

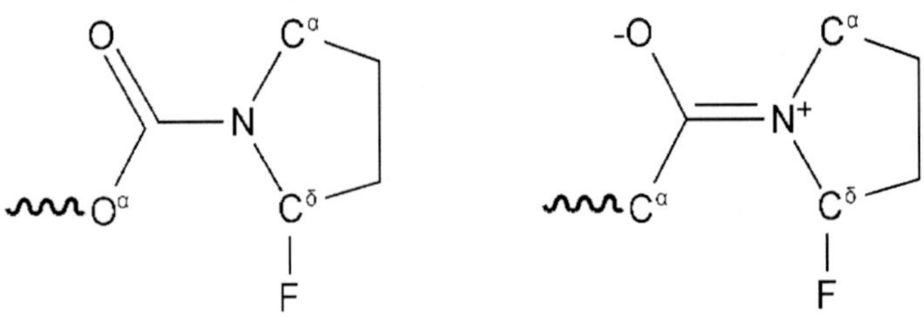

Figure: *An electronegative substituent near the amide nitrogen favors the single-bonded resonance form (left) over the double-bonded form (right).*

A peptide bond can be broken by amide hydrolysis (the adding of water). The peptide bonds in proteins are metastable, meaning that in the presence of water they will break spontaneously, releasing 2-4 kcal/mol of free energy, but this process is extremely slow.

In living organisms, the process is facilitated by enzymes. Living organisms also employ enzymes to form peptide bonds; this process requires free energy. The wavelength of absorbance for a peptide bond is 190-230 nm.

Resonance Forms of the Peptide Group

The amide group has two resonance forms, which confer several important properties. First, it stabilizes the group by roughly 20 kcal/mol, making it less reactive than many similar groups (such as esters). The resonance suggests that the amide group has a partial double bond character, estimated at 40% under typical conditions. The peptide group is uncharged at all normal pH values, but its double-bonded resonance form gives it an unusually large dipole moment, roughly 3.5 Debye (0.7 electron-angstrom). These dipole moments can line up in certain secondary structures (such as the α-helix), producing a large net dipole.

The partial double bond character can be strengthened or weakened by modifications that favour one resonance form over another. For example, the double-bonded form is disfavored in hydrophobic environments, because of its charge. Conversely, donating a hydrogen bond to the amide oxygen or accepting a hydrogen bond from the amide nitrogen should favour the double-bonded form, because the hydrogen bond should be stronger to the charged form than to the uncharged, single-bonded form.

By contrast, donating a hydrogen bond to an amide nitrogen in an X-Pro peptide bond should favour the single-bonded form; donating it to the double-bonded form would give the nitrogen five quasi-covalent bonds! Similarly, a strongly electronegative substituent (such as fluorine) near the amide nitrogen favors the single-bonded form, by competing with the amide oxygen to "steal" an electron from the amide nitrogen.

Cis/trans Isomers of the Peptide Group

The partial double bond renders the amide group planar, occurring in either the cis or trans isomers. In the unfolded state of proteins, the peptide groups are free to isomerize and adopt both isomers; however, in the folded state, only a single isomer is adopted at each position (with rare exceptions). The trans form is preferred overwhelmingly in most peptide bonds (roughly 1000:1 ratio in trans:cis populations). However, X-Pro peptide groups tend to have a roughly 3:1 ratio, presumably because the symmetry between the C^{α} and C^{δ} atoms of proline makes the cis and trans isomers nearly equal in energy.

The dihedral angle associated with the peptide group (defined by the four atoms $c^{\alpha} - C' - N - C^{\alpha}$) is denoted ω; ω = 0° for the cis isomer and ω = 180° for the trans isomer. Amide groups can isomerize about the C-N bond between the cis and trans forms, albeit slowly τ ~ (20 seconds at room temperature). The transition states ω = ± 90° requires that the partial double bond be broken, so that the activation energy is roughly 20 kcal/mol. However, the activation energy can be lowered (and the isomerization catalysed) by changes that favour the single-bonded form, such as placing the peptide group in a hydrophobic environment or donating a hydrogen bond to the nitrogen atom of an X-Pro peptide group. Both of these mechanisms for lowering the activation energy have been observed in peptidyl prolyl isomerases (PPIases), which are naturally occurring enzymes that catalyse the cis-trans isomerization of X-Pro peptide bonds.

Conformational protein folding is usually much faster (typically 10-100 ms) than cis-trans isomerization (10-100 s). A nonnative isomer of some peptide groups can disrupt the conformational folding significantly, either slowing it or preventing it from even occurring until the native isomer is reached. However, not all peptide groups have the same effect on folding; nonnative isomers of other peptide groups may not affect folding at all.

Chemical Reactions

Due to its resonance stabilization, the peptide bond is relatively unreactive under physiological conditions, even less than similar compounds such as esters. Nevertheless, peptide bonds can undergo chemical reactions, usually through an attack of an electronegative atom on the carbonyl carbon, breaking the carbonyl double bond and forming a tetrahedral intermediate. This is the pathway followed in proteolysis and, more generally, in N-O acyl exchange reactions such as those of inteins. When the functional group attacking the peptide bond is a thiol, hydroxyl or amine, the resulting molecule may be called a cyclol or, more specifically, a thiacyclol, an oxacyclol or an azacyclol, respectively.

7

Biochemistry of Genome

In modern molecular biology and genetics, the genome is the entirety of an organism's hereditary information. It is encoded either in DNA or, for many types of virus, in RNA. The genome includes both the genes and the non-coding sequences of the DNA.

Origin of Term

The term was adapted in 1920 by Hans Winkler, Professor of Botany at the University of Hamburg, Germany. In Greek, the word *genome* means I become, I am born, to come into being. The Oxford English Dictionary suggests the name to be a blend of the words *gene* and *chromosome*. A few related -*ome* words already existed, such as *biome* and *rhizome*, forming a vocabulary into which *genome* fits systematically.

Overview

Some organisms have multiple copies of chromosomes, diploid, triploid, tetraploid and so on. In classical genetics, in a sexually reproducing organism (typically eukarya) the gamete has half of the number of chromosome of the somatic cell and the genome is a full set of chromosomes in a gamete. In haploid organisms, including cells of bacteria, archaea, and in organelles including mitochondria and chloroplasts, or viruses, that similarly contain genes, the single or set of circular and/or linear chains of DNA (or RNA for some viruses), likewise constitute the *genome*. The term genome can be applied specifically to mean that stored on a complete set of *nuclear DNA* (i.e., the "nuclear genome") but can also be applied to that stored within organelles that contain their own DNA, as with the "mitochondrial genome" or the "chloroplast genome". Additionally, the genome can comprise nonchromosomal genetic elements such as viruses, plasmids, and transposable elements.

When people say that the genome of a sexually reproducing species has been "sequenced", typically they are referring to a determination of the sequences of one set of autosomes and one of each type of sex chromosome, which together represent both of the possible sexes. Even in species that exist in only one sex, what is described as "a genome sequence" may be a composite read from the chromosomes of various individuals. In general use, the phrase "genetic makeup" is sometimes used conversationally to mean the genome of a particular individual or organism.

The study of the global properties of genomes of related organisms is usually referred to as genomics, which distinguishes it from genetics which generally studies the properties of single genes or groups of genes.

Both the number of base pairs and the number of genes vary widely from one species to another, and there is only a rough correlation between the two (an observation known as the C-value paradox). At present, the highest known number of genes is around 60,000, for the protozoan causing trichomoniasis, almost three times as many as in the human genome.

An analogy to the human genome stored on DNA is that of instructions stored in a library:

- The library would contain 46 books (chromosomes)
- The books range in size from 400 to 3340 pages (genes)
- which is 48 to 250 million letters (A,C,G,T) per book.
- Hence the library contains over six billion letters total;
- The library fits into a cell nucleus the size of a pinpoint;
- A copy of the library (all 46 books) is contained in almost every cell of our body.

Types

Most biological entities that are more complex than a virus sometimes or always carry additional genetic material besides that which resides in their chromosomes. In some contexts, such as sequencing the genome of a pathogenic microbe, "genome" is meant to include information stored on this auxiliary material, which is carried in plasmids. In such circumstances then, "genome" describes all of the genes and information on non-coding DNA that have the potential to be present. In eukaryotes such as plants, protozoa and animals, however, "genome" carries the typical connotation of only information on chromosomal DNA. So although these organisms contain chloroplasts and/or mitochondria that have their own DNA, the genetic information contained by DNA within these organelles is not considered part of the genome. In fact, mitochondria are sometimes said to have their own genome often referred to as the "mitochondrial genome". The DNA found within the chloroplast may be referred to as the "plastome".

Genomes and Genetic Variation

Note that a genome does not capture the genetic diversity or the genetic polymorphism of a species. For example, the human genome sequence in principle could be determined from just half the information on the DNA of one cell from one individual. To learn what variations in genetic information underlie particular traits or diseases requires comparisons across individuals. This point explains the common usage of "genome" (which parallels a common usage of "gene") to refer not to the information in any particular DNA sequence, but to a whole family of sequences that share a biological context.

Although this concept may seem counter intuitive, it is the same concept that says there is no particular shape that is the shape of a cheetah. Cheetahs vary, and so do the sequences of their genomes. Yet both the individual animals and their sequences share commonalities, so one can learn something about cheetahs and "cheetah-ness" from a single example of either.

Sequencing and Mapping

The Human Genome Project was organized to map and to sequence the human genome. Other genome projects include mouse, rice, the plant *Arabidopsis thaliana*, the puffer fish, bacteria like E. coli, etc. In 1976, Walter Fiers at the University of Ghent (Belgium) was the first to establish the complete nucleotide sequence of a viral RNA-genome (bacteriophage MS2). The first DNA-genome project to be completed was the Phage Φ-X174, with only 5386 base pairs, which was sequenced by Fred Sanger in 1977. The first bacterial genome to be completed was that of Haemophilus influenzae, completed by a team at The Institute for Genomic Research in 1995.

The development of new technologies has dramatically decreased the difficulty and cost of sequencing, and the number of complete genome sequences is rising rapidly. Among many genome database sites, the one maintained by the US National Institutes of Health is inclusive. These new technologies open up the prospect of personal genome sequencing as an important diagnostic tool. A major step toward that goal was the May 2007 *New York Times* announcement that the full genome of DNA pioneer James D. Watson was deciphered.

Whereas a genome sequence lists the order of every DNA base in a genome, a genome map identifies the landmarks. A genome map is less detailed than a genome sequence and aids in navigating around the genome.

Genome Evolution

Genomes are more than the sum of an organism's genes and have traits that may be measured and studied without reference to the details of any particular genes and their products. Researchers compare traits such as *chromosome number* (karyotype), genome size, gene order, codon usage bias, and GC-content to determine what mechanisms could have produced the great variety of genomes that exist today.

Duplications play a major role in shaping the genome. Duplications may range from extension of short tandem repeats, to duplication of a cluster of genes, and all the way to duplications of entire chromosomes or even entire genomes. Such duplications are probably fundamental to the creation of genetic novelty. Horizontal gene transfer is invoked to explain how there is often extreme similarity between small portions of the genomes of two organisms that are otherwise very distantly related. Horizontal gene transfer seems to be common among many microbes. Also, eukaryotic cells seem to have experienced a transfer of some genetic material from their chloroplast and mitochondrial genomes to their nuclear chromosomes.

Basics of Biological Chemistry

Biochemistry, sometimes abbreviated as "BioChem", is the study of chemical processes in living organisms. Biochemistry governs all living organisms and living processes. By controlling information flow through biochemical signalling and the flow of chemical energy through metabolism; biochemical processes give rise to the seemingly magical phenomenon of life. Much of biochemistry deals with the structures and functions of cellular components such as proteins, carbohydrates, lipids, nucleic acids and other biomolecules although increasingly processes rather than individual molecules are the main focus. Over the last 40 years biochemistry has become so successful at explaining living processes that now almost all areas of the life sciences from botany to medicine are engaged in biochemical research. Today the main focus of pure biochemistry is in understanding how biological molecules give rise to the processes that occur within living cells which in turn relates greatly to the study and understanding of whole organisms.

Among the vast number of different biomolecules, many are complex and large molecules (called *polymers*), which are composed of similar repeating subunits (called *monomers*). Each class of polymeric biomolecule has a different set of subunit types. For example, a protein is a polymer whose subunits are selected from a set of 20 or more amino acids. Biochemistry studies the chemical properties of important biological molecules, like proteins, and in particular the chemistry of enzyme-catalysed reactions. The biochemistry of cell metabolism and the endocrine system has been extensively described. Other areas of biochemistry include the genetic code (DNA, RNA), protein synthesis, cell membrane transport, and signal transduction.

History

Originally, it was generally believed that life was not subject to the laws of science the way non-life was. It was thought that only living beings could produce the molecules of life (from other, previously existing biomolecules). Then, in 1828, Friedrich Wöhler published a paper on the synthesis of urea, proving that organic compounds can be created artificially.

The dawn of biochemistry may have been the discovery of the first enzyme, diastase (today called amylase), in 1833 by Anselme Payen. Eduard Buchner contributed the first demonstration of a complex biochemical process outside of a cell in 1896: alcoholic fermentation in cell extracts of yeast. Although the term "biochemistry" seems to have been first used in 1882, it is generally accepted that the formal coinage of biochemistry occurred in 1903 by Carl Neuberg, a German chemist. Previously, this area would have been referred to as physiological chemistry.

Since then, biochemistry has advanced, especially since the mid-20th century, with the development of new techniques such as chromatography, X-ray diffraction, dual polarisation interferometry, NMR spectroscopy, radioisotopic labelling, electron microscopy and molecular dynamics simulations. These techniques allowed for the discovery and detailed analysis of many molecules and metabolic pathways of the cell, such as glycolysis and the Krebs cycle (citric acid cycle).

Another significant historic event in biochemistry is the discovery of the gene and its role in the transfer of information in the cell. This part of biochemistry is often called molecular biology. In the 1950s, James D. Watson, Francis Crick, Rosalind Franklin, and Maurice Wilkins were instrumental in solving DNA structure and suggesting its relationship with genetic transfer of information. In 1958, George Beadle and Edward Tatum received the Nobel Prize for work in fungi showing that one gene produces one enzyme. In 1988, Colin Pitchfork was the first person convicted of murder with DNA evidence, which led to growth of forensic science. More recently, Andrew Z. Fire and Craig C. Mello received the 2006 Nobel Prize for discovering the role of RNA interference (RNAi), in the silencing of gene expression

Today, there are three main types of biochemistry. Plant biochemistry involves the study of the biochemistry of autotrophic organisms such as photosynthesis and other plant specific biochemical processes. General biochemistry encompasses both plant and animal biochemistry. Human/medical/medicinal biochemistry focuses on the biochemistry of humans and medical illnesses.

Monomers and Polymers

The four main classes of molecules in biochemistry are carbohydrates, lipids, proteins, and nucleic acids. Many biological molecules are polymers: in this terminology, *monomers* are relatively small micromolecules that are linked together to create large macromolecules, which are known as *polymers*. When monomers are linked together to synthesize a biological polymer, they undergo a process called dehydration synthesis.

Carbohydrates

Carbohydrates are made from monomers called *monosaccharides*. Some of these monosaccharides include glucose ($C_6H_{12}O_6$), fructose ($C_6H_{12}O_6$), and deoxyribose ($C_5H_{10}O_4$). When two monosaccharides undergo dehydration synthesis, water is produced, as two hydrogen atoms and one oxygen atom are lost from the two monosaccharides' hydroxyl group.

Lipids

Lipids are usually made from one molecule of glycerol combined with other molecules. In triglycerides, the main group of bulk lipids, there is one molecule of glycerol and three fatty acids. Fatty acids are considered the monomer in that case, and may be saturated (no double bonds in the carbon chain) or unsaturated (one or more double bonds in the carbon chain). Lipids, especially phospholipids, are also used in various pharmaceutical products, either as co-solubilisers (e.g. in parenteral infusions) or else as drug carrier components (e.g. in a liposome or transfersome).

Proteins

Proteins are very large molecules – macro-biopolymers – made from monomers called *amino acids*. There are 20 standard amino acids, each containing a carboxyl group, an amino group, and a side chain (known as an "R" group). The "R" group is what makes each amino acid different, and the properties of the side chains greatly influence the overall three-dimensional conformation of a protein. When amino acids combine, they form a special bond called a peptide bond through dehydration synthesis, and become a *polypeptide*, or protein.

Nucleic Acids

Nucleic acids are the molecules that make up DNA, an extremely important substance which all cellular organisms use to store their genetic information. The most common nucleic acids are deoxyribonucleic acid and ribonucleic acid. Their monomers are called nucleotides. The most common nucleotides are adenine, cytosine, guanine, thymine, and uracil. Adenine binds with thymine and uracil; thymine only binds with adenine; and cytosine and guanine can only bind with each other.

Carbohydrates

The function of carbohydrates includes energy storage and providing structure. Sugars are carbohydrates, but not all carbohydrates are sugars. There are more carbohydrates on Earth than any other known type of biomolecule; they are used to store energy and genetic information, as well as play important roles in cell to cell interactions and communications.

Monosaccharides

The simplest type of carbohydrate is a monosaccharide, which among other properties contains carbon, hydrogen, and oxygen, mostly in a ratio of 1:2:1 (generalized formula $C_nH_{2n}O_n$, where n is at least 3). Glucose, one of the most important carbohydrates, is an example of a monosaccharide. So is fructose, the sugar commonly associated with the sweet taste of fruits.

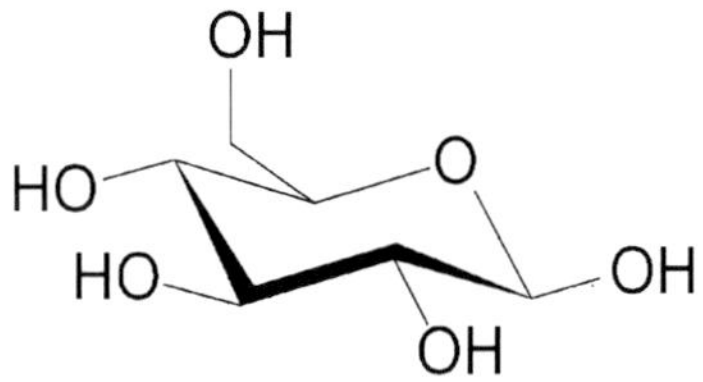

Figure: *Glucose*

Some carbohydrates (especially after condensation to oligo- and polysaccharides) contain less carbon relative to H and O, which still are present in 2:1 (H:O) ratio. Monosaccharides can be grouped into aldoses (having an aldehyde group at the end of the chain, e.g. glucose) and ketoses (having a keto group in their chain; e.g. fructose).

Both aldoses and ketoses occur in an equilibrium (starting with chain lengths of C4) cyclic forms. These are generated by bond formation between one of the hydroxyl groups of the sugar chain with the carbon of the aldehyde or keto group to form a hemiacetal bond. This leads to saturated five-membered (in furanoses) or six-membered (in pyranoses) heterocyclic rings containing one O as heteroatom.

Disaccharides

Two monosaccharides can be joined together using dehydration synthesis, in which a hydrogen atom is removed from the end of one molecule and a hydroxyl group (—OH) is removed from the other; the remaining residues are then attached at the sites from which the atoms were removed. The H—OH or H_2O is then released as a molecule of water, hence the term *dehydration*. The new molecule, consisting of two monosaccharides, is called a *disaccharide* and is conjoined together by a glycosidic or ether bond. The reverse reaction can also occur, using a molecule of water to split up a disaccharide and break the glycosidic bond; this is termed *hydrolysis*. The most well-known disaccharide is sucrose, ordinary sugar (in scientific contexts, called *table sugar* or *cane sugar* to differentiate it from other sugars). Sucrose consists of a glucose molecule and a fructose molecule joined together. Another important disaccharide is lactose, consisting of a glucose molecule and a galactose molecule. As most humans age, the production of lactase, the enzyme that hydrolyzes lactose back into glucose and galactose, typically decreases. This results in lactase deficiency, also called *lactose intolerance*.

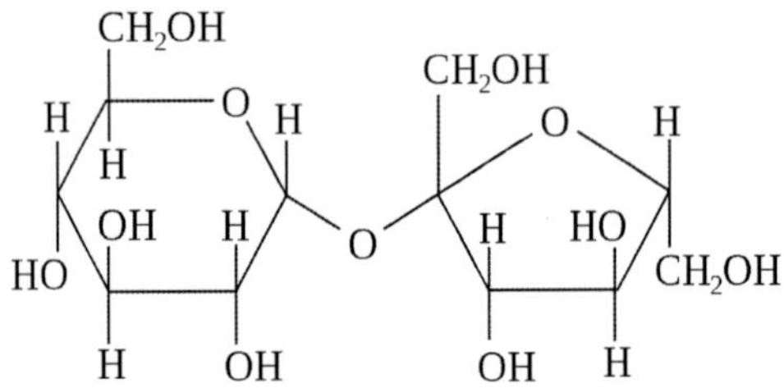

Figure: *ordinary sugar and probably the most familiar carbohydrate.*

Sugar polymers are characterised by having reducing or non-reducing ends. A reducing end of a carbohydrate is a carbon atom which can be in equilibrium with the open-chain aldehyde or keto form.

If the joining of monomers takes place at such a carbon atom, the free hydroxy group of the pyranose or furanose form is exchanged with an OH-side chain of another sugar, yielding a full acetal.

This prevents opening of the chain to the aldehyde or keto form and renders the modified residue non-reducing. Lactose contains a reducing end at its glucose moiety, whereas the galactose moiety form a full acetal with the C4-OH group of glucose. Saccharose does not have a reducing end because of full acetal formation between the aldehyde carbon of glucose (C1) and the keto carbon of fructose (C2).

Oligosaccharides and Polysaccharides

When a few (around three to six) monosaccharides are joined together, it is called an *oligosaccharide*. These molecules tend to be used as markers and signals, as well as having some other uses.

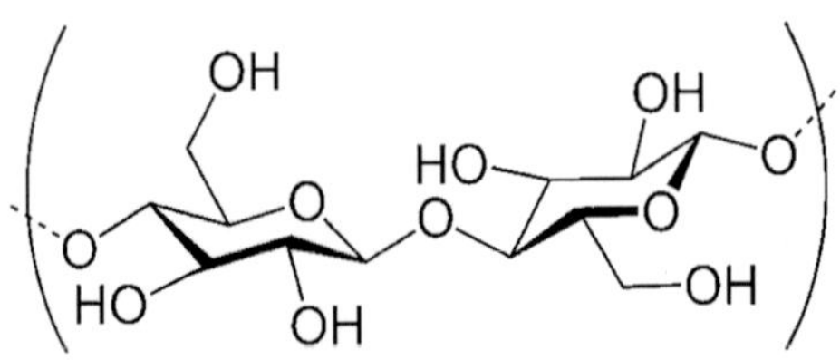

Figure: *Cellulose as polymer of β-D-glucose*

Many monosaccharides joined together make a polysaccharide. They can be joined together in one long linear chain, or they may be branched. Two of the most common polysaccharides are cellulose and glycogen, both consisting of repeating glucose monomers.

- *Cellulose* is made by plants and is an important structural component of their cell walls. Humans can neither manufacture nor digest it.
- *Glycogen*, on the other hand, is an animal carbohydrate; humans and other animals use it as a form of energy storage.

Use of Carbohydrates as an Energy Source

Glucose is the major energy source in most life forms. For instance, polysaccharides are broken down into their monomers (glycogen phosphorylase removes glucose residues from glycogen). Disaccharides like lactose or sucrose are cleaved into their two component monosaccharides.

Glycolysis (Anaerobic)

Glucose is mainly metabolized by a very important ten-step pathway called glycolysis, the net result of which is to break down one molecule of glucose into two molecules of pyruvate; this also produces a net two molecules of ATP, the energy currency of cells, along with two reducing equivalents in the form of converting NAD^+ to NADH. This does not require oxygen; if no oxygen is available (or the cell cannot use oxygen), the NAD is restored by converting the pyruvate to lactate (lactic acid) (e.g. in humans) or to ethanol plus carbon dioxide (e.g. in yeast). Other monosaccharides like galactose and fructose can be converted into intermediates of the glycolytic pathway.

Aerobic

In aerobic cells with sufficient oxygen, like most human cells, the pyruvate is further metabolized. It is irreversibly converted to acetyl-CoA, giving off one carbon atom as the waste product carbon dioxide, generating another reducing equivalent as NADH. The two molecules acetyl-CoA (from one molecule of glucose) then enter the citric acid cycle, producing two more molecules of ATP, six more NADH molecules and two reduced (ubi)quinones (via $FADH_2$ as enzyme-bound cofactor), and releasing the remaining carbon atoms as carbon dioxide.

The produced NADH and quinol molecules then feed into the enzyme complexes of the respiratory chain, an electron transport system transferring the electrons ultimately to oxygen and conserving the released energy in the form of a proton gradient over a membrane (inner mitochondrial membrane in eukaryotes). Thereby, oxygen is reduced to water and the original electron acceptors NAD^+ and quinone are regenerated. This is why humans breathe in oxygen and breathe out carbon dioxide. The energy released from transferring the electrons from high-energy states in NADH and quinol is conserved first as proton gradient and converted to ATP via ATP synthase. This generates an additional *28* molecules of ATP (24 from the 8 NADH + 4 from the 2 quinols), totalling to 32 molecules of ATP conserved per degraded glucose (two from glycolysis + two from the citrate cycle). It is clear that using oxygen to completely oxidize glucose provides an organism with far more energy than any oxygen-independent metabolic feature, and this is thought to be the reason why complex life appeared only after Earth's atmosphere accumulated large amounts of oxygen.

Gluconeogenesis

In vertebrates, vigorously contracting skeletal muscles (during weightlifting or sprinting, for example) do not receive enough oxygen to meet the energy demand, and so they shift to anaerobic metabolism, converting glucose to lactate. The liver regenerates the glucose, using a process called gluconeogenesis.

This process is not quite the opposite of glycolysis, and actually requires three times the amount of energy gained from glycolysis (six molecules of ATP are used, compared to the two gained in glycolysis). Analogous to the above reactions, the glucose produced can then undergo glycolysis in tissues that need energy, be stored as glycogen (or starch in plants), or be converted to other monosaccharides or joined into di- or oligosaccharides. The combined pathways of glycolysis during exercise, lactate's crossing via the bloodstream to the liver, subsequent gluconeogenesis and release of glucose into the bloodstream is called the Cori cycle.

Proteins

Like carbohydrates, some proteins perform largely structural roles. For instance, movements of the proteins actin and myosin ultimately are responsible for the contraction of skeletal muscle. One property many proteins have is that they specifically bind to a certain molecule or class of molecules—they may be *extremely* selective in what they bind. Antibodies are an example of proteins that attach to one specific type of molecule.

In fact, the enzyme-linked immunosorbent assay (ELISA), which uses antibodies, is currently one of the most sensitive tests modern medicine uses to detect various biomolecules. Probably the most important proteins, however, are the enzymes. These molecules recognize specific reactant molecules called *substrates*; they then catalyse the reaction between them. By lowering the activation energy, the enzyme speeds up that reaction by a rate of ^{11}or more: a reaction that would normally take over 3,000 years to complete spontaneously might take less than a second with an enzyme. The enzyme itself is not used up in the process, and is free to catalyse the same reaction with a new set of substrates. Using various modifiers, the activity of the enzyme can be regulated, enabling control of the biochemistry of the cell as a whole.

In essence, proteins are chains of amino acids. An amino acid consists of a carbon atom bound to four groups. One is an amino group, $—NH_2$, and one is a carboxylic acid group, —COOH (although these exist as $—NH_3^+$ and —COO^- under physiologic conditions). The third is a simple hydrogen atom. The fourth is commonly denoted "—R" and is different for each amino acid. There are twenty standard amino acids. Some of these have functions by themselves or in a modified form; for instance, glutamate functions as an important neurotransmitter.

Amino acids can be joined together via a peptide bond. In this dehydration synthesis, a water molecule is removed and the peptide bond connects the nitrogen of one amino acid's amino group to the carbon of the other's carboxylic acid group. The resulting molecule is called a *dipeptide*, and short stretches of amino acids (usually, fewer than around thirty) are called *peptides* or polypeptides. Longer stretches merit the title *proteins*. As an example, the important blood serum protein albumin contains 585 amino acid residues.

The structure of proteins is traditionally described in a hierarchy of four levels. The primary structure of a protein simply consists of its linear sequence of amino acids; for instance, "alanine-glycine-tryptophan-serine-glutamate-asparagine-glycine-lysine-...". Secondary structure is concerned with local morphology (morphology being the study of structure).

Some combinations of amino acids will tend to curl up in a coil called an α-helix or into a sheet called a β-sheet; some α-helixes can be seen in the hemoglobin schematic above. Tertiary structure is the entire three-dimensional shape of the protein.

This shape is determined by the sequence of amino acids. In fact, a single change can change the entire structure. The alpha chain of hemoglobin contains 146 amino acid residues; substitution of the glutamate residue at position 6 with a valine residue changes the behaviour of hemoglobin so much that it results in sickle-cell disease. Finally quaternary structure is concerned with the structure of a protein with multiple peptide subunits, like hemoglobin with its four subunits. Not all proteins have more than one subunit.

Ingested proteins are usually broken up into single amino acids or dipeptides in the small intestine, and then absorbed. They can then be joined together to make new proteins. Intermediate products of glycolysis, the citric acid cycle, and the pentose phosphate pathway can be used to make all twenty amino acids, and most bacteria and plants possess all the necessary enzymes to synthesize them.

Humans and other mammals, however, can only synthesize half of them. They cannot synthesize isoleucine, leucine, lysine, methionine, phenylalanine, threonine, tryptophan, and valine. These are the essential amino acids, since it is essential to ingest them. Mammals do possess the enzymes to synthesize alanine, asparagine, aspartate, cysteine, glutamate, glutamine, glycine, proline, serine, and tyrosine, the nonessential amino acids. While they can synthesize arginine and histidine, they cannot produce it in sufficient amounts for young, growing animals, and so these are often considered essential amino acids.

If the amino group is removed from an amino acid, it leaves behind a carbon skeleton called an α-keto acid. Enzymes called transaminases can easily transfer the amino group from one amino acid (making it an α-keto acid) to another α-keto acid (making it an amino acid). This is important in the biosynthesis of amino acids, as for many of the pathways, intermediates from other biochemical pathways are converted to the α-keto acid skeleton, and then an amino group is added, often via transamination. The amino acids may then be linked together to make a protein.

A similar process is used to break down proteins. It is first hydrolysed into its component amino acids. Free ammonia (NH_3), existing as the ammonium ion (NH_4^+) in blood, is toxic to life forms. A suitable method for excreting it must therefore exist. Different strategies have evolved in different animals, depending on the animals' needs. Unicellular organisms, of course, simply release the ammonia into the environment. Similarly, bony fish can release the ammonia into the water where it is quickly diluted. In general, mammals convert the ammonia into urea, via the urea cycle.

Lipids

The term lipid comprises a diverse range of molecules and to some extent is a catchall for relatively water-insoluble or nonpolar compounds of biological origin, including waxes, fatty acids, fatty-acid derived phospholipids, sphingolipids, glycolipids and terpenoids (e.g. retinoids and steroids). Some lipids are linear aliphatic molecules, while others have ring structures. Some are aromatic, while others are not. Some are flexible, while others are rigid.

Most lipids have some polar character in addition to being largely nonpolar. Generally, the bulk of their structure is nonpolar or hydrophobic ("water-fearing"), meaning that it does not interact well with polar solvents like water. Another part of their structure is polar or hydrophilic ("water-loving") and will tend to associate with polar solvents like water.

This makes them amphiphilic molecules (having both hydrophobic and hydrophilic portions). In the case of cholesterol, the polar group is a mere -

OH (hydroxyl or alcohol). In the case of phospholipids, the polar groups are considerably larger and more polar, as described below.

Lipids are an integral part of our daily diet. Most oils and milk products that we use for cooking and eating like butter, cheese, ghee etc., are composed of fats. Vegetable oils are rich in various polyunsaturated fatty acids (PUFA). Lipid-containing foods undergo digestion within the body and are broken into fatty acids and glycerol, which are the final degradation products of fats and lipids.

Nucleic Acids

A nucleic acid is a complex, high-molecular-weight biochemical macromolecule composed of nucleotide chains that convey genetic information. The most common nucleic acids are deoxyribonucleic acid (DNA) and ribonucleic acid (RNA). Nucleic acids are found in all living cells and viruses. Aside from the genetic material of the cell, nucleic acids often play a role as second messengers, as well as forming the base molecule for adenosine triphosphate, the primary energy-carrier molecule found in all living organisms.

Nucleic acid, so called because of its prevalence in cellular nuclei, is the generic name of the family of biopolymers. The monomers are called nucleotides, and each consists of three components: a nitrogenous heterocyclic base (either a purine or a pyrimidine), a pentose sugar, and a phosphate group.

Different nucleic acid types differ in the specific sugar found in their chain (e.g. DNA or deoxyribonucleic acid contains 2-deoxyriboses). Also, the nitrogenous bases possible in the two nucleic acids are different: adenine, cytosine, and guanine occur in both RNA and DNA, while thymine occurs only in DNA and uracil occurs in RNA.

Relationship to other "Molecular-scale" Biological Sciences

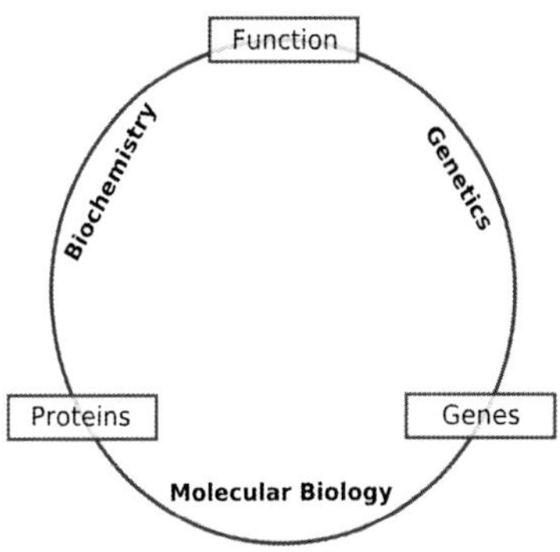

Figure: Schematic relationship between biochemistry, genetics and molecular biology

Researchers in biochemistry use specific techniques native to biochemistry, but increasingly combine these with techniques and ideas from genetics, molecular biology and biophysics. There has never been a hard-line between these disciplines in terms of content and technique. Today the terms *molecular biology* and *biochemistry* are nearly interchangeable. The schematic that depicts one possible view of the relationship between the fields:

- *Biochemistry* is the study of the chemical substances and vital processes occurring in living organisms. Biochemists focus heavily on the role, function, and structure of biomolecules. The study of the chemistry behind biological processes and the synthesis of biologically active molecules are examples of biochemistry.
- *Genetics* is the study of the effect of genetic differences on organisms. Often this can be inferred by the absence of a normal component (e.g. one gene). The study of "mutants" – organisms which lack one or more functional components with respect to the so-called "wild type" or normal phenotype. Genetic interactions (epistasis) can often confound simple interpretations of such "knock-out" studies.
- *Molecular biology* is the study of molecular underpinnings of the process of replication, transcription and translation of the genetic material. The central dogma of molecular biology where genetic material is transcribed into RNA and then translated into protein, despite being an oversimplified picture of molecular biology, still provides a good starting point for understanding the field. This picture, however, is undergoing revision in light of emerging novel roles for RNA.
- *Chemical Biology* seeks to develop new tools based on small molecules that allow minimal perturbation of biological systems while providing detailed information about their function. Further, chemical biology employs biological systems to create non-natural hybrids between biomolecules and synthetic devices (for example emptied viral capsids that can deliver gene therapy or drug molecules).

Bibliography

Ambrose, E. J., and Easty, D. M.: *Cell Biology: Reading*, Addison- Wesley, Mass, 1970.

Barrington, E. J. W.: *Biochemistry of Primitive Deuterostomians*, London, Academic Press, 1974.

Bauer, M.W., and G. Gaskell : *Biotechnology: The Making of a Global Controversy*, Cambridge: Cambridge University Press, 2002.

Bowler, Peter J.: *The Mendelian Revolution: The Emergence of Hereditarian Concepts in Modern Science and Society*. Baltimore: Johns Hopkins University Press, 1989.

Broach, J.R. : *The Molecular Biology of the Yeast,* Cold Spring Harbor: Cold Spring Harbor Laboratory, 1981.

Carroll SB, Grenier J, Weatherbee SD: *From DNA to Diversity: Molecular Genetics and the Evolution of Animal Design*. Oxford: Blackwell Publishing, 2005.

Chauhan, B.S. : *Principles of Biochemistry and Biophysics,* Laxmi Publications, Delhi, 2008.

Dhruve, J J : *Biochemistry at a Glance*, Agrotech Pub, Delhi, 2007.

Friedberg, E.C. : *DNA Repair,* New York: WH Freeman and Company, 1985.

Ganguly, Smriti : *Biochemistry of Biomolecules,* Pearl Books, Delhi, 2007.

Harris, John : *Wonderwoman and Superman: The Ethics of Human Biotechnology,* Oxford: Oxford University Press, 1992.

Hartl D, Jones E: *Genetics: Analysis of Genes and Genomes*. Jones & Bartlett, 2005.

Kitcher, Philip : *The Lives to Come: The Genetic Revolution and Human Possibilities*, New York: Simon and Schuster, 1996.

Lata Bhattacharya: *Biochemistry of Nutrition*, Discovery, Delhi, 2010.

Lodish, Harvey : *Molecular Cell Biology*, New York: W. H. Freeman, 2000.

Morholt, E. and Brandwein, P.F. : *Sourcebook for the Biological Sciences,* San Diego: Harcourt Brace Jovanovich, 1986.

Morris, Jonathan : *Ethics and Biotechnology*, New York: Chelsea House, 2005.

Narvekar, Raghunath : *Molecular Biochemistry : Principles and Practices,* Adhyayan Pub, Delhi, 2008.

Patel, S V ; U K Kandoliya: *Biochemistry at a Glance*, Agrotech Pub, Delhi, 2007.

Primrose, S. B. : *Molecular Biotechnology,* Boston: Blackwell Scientific Publications, 1991.

Reuter, Lars : *Modern Biotechnology in Postmodern Times? A Reflection on European Policies and Human Agency,* Dordrecht, Netherlands: Springer, 2003.

Sarath Chandra Bose: *Biochemistry : A Practical Manual*, Pharma Med Press, Delhi, 2010.

Sharma, Pradeep: *Biochemistry and Organisation of Cells*, RBSA Pub, Delhi, 2006.

Singh, Bimla: *Biochemistry and Genomic Revolution*, Vista International Pub, Delhi, 2006.

Trevor H. : *Transforming Matter: A History of Chemistry from Alchemy to the Buckyball,* Baltimore: Johns Hopkins University Press, 2001.

William Alec : *The Chemical Industry*, London: Longman Group Limited, 1971.

Yadav, Neelam: *Biochemistry and Biotechnology : A Laboratory Manual*, Pointer, Delhi, 2007.

Index

❑❑❑